A HUMAN ERROR APPROACH
ACCIDENT ANALYSIS

A Human Error Approach to Aviation Accident Analysis

The Human Factors Analysis and Classification System

DOUGLAS A. WIEGMANN
University of Illinois at Urbana-Champaign
SCOTT A. SHAPPELL
Civil Aerospace Medical Institute

ASHGATE

© Douglas A. Wiegmann and Scott A. Shappell 2003

All rights reserved. No part of this publication may be reproduced, stored in a retrieval system, or transmitted in any form or by any means, electronic, mechanical, photocopying, recording or otherwise without the prior permission of the publisher.

The authors have asserted their moral right under the Copyright, Designs and Patents Act, 1988, to be identified as the authors of this work.

Published by
Ashgate Publishing Limited
Wey Court East
Union Street
Farnham
Surrey GU9 7PT
England

Ashgate Publishing Company
110 Cherry Street
Suite 3-1
Burlington, VT 05401-3818
USA

Ashgate website: http://www.ashgate.com

British Library Cataloguing in Publication Data
Wiegmann, Douglas A.
 A human error approach to aviation accident analysis
 1. Aircraft accidents - Human factors 2. Aircraft accidents
 Investigation Methodology
 I. Title II. Shappell, Scott A.
 363.1'2414

Library of Congress Cataloging-in-Publication Data
Wiegmann, Douglas
 A. A human error approach to aviation accident analysis : the human factors analysis and classification system / Douglas A. Wiegmann, and Scott A. Shappell.
  p. cm.
 Includes bibliographical references and index.
 ISBN 978-0-7546-1875-1 (alk. paper) —ISBN 978-0-7546-1873-7 (pbk. : alk. paper)
  1. Aircraft accidents—Investigation. 2. Aeronautics—Human factors. I. Shappell, Scott A. II. Title.

TL553.5.W54 2003
363.12'465—dc21

2003048192

ISBN 978 0 7546 1875 1 (hbk)
ISBN 978 0 7546 1873 7 (pbk)

Reprinted 2004, 2005, 2006, 2009 (twice), 2011, 2013, 2014

Printed in the United Kingdom by Henry Ling Limited, at the Dorset Press, Dorchester, DT1 1HD

Contents

List of Figures — vii
List of Tables — x
Acknowledgements — xi
Preface — xii

1 *Errare Humanum Est* – To Err is Human — 1
   Aviation Safety Trends — 3
   Some Reasons for Concern — 8
   Human Error and Aviation Accidents — 10
   Engineering Aspects of an Investigation — 12
   Human Factors Aspects of an Investigation — 15
   Conclusion — 18

2 Human Error Perspectives — 20
   The Cognitive Perspective — 21
   The Ergonomic Perspective — 26
   The Behavioral Perspective — 30
   The Aeromedical Perspective — 32
   The Psychosocial Perspective — 34
   The Organizational Perspective — 37
   Conclusion — 44

3 The Human Factors Analysis and Classification System (HFACS) — 45
   Reason's Model of Accident Causation — 45
      Elements of a Productive System — 45
      Breakdown of a Productive System — 47
   Strengths and Limitations of Reason's Model — 49
   Defining the Holes in the Cheese — 50
   Unsafe Acts of Operators — 50
      Errors — 51
      Violations — 55
   Preconditions for Unsafe Acts — 56
      Condition of Operators — 57
      Personnel Factors — 60
      Environmental Factors — 61

	Unsafe Supervision	63
	Organizational influences	66
	Conclusion	70
4	Aviation Case Studies using HFACS	72
	Sometimes Experience does Count	73
	Human Factors Analysis using HFACS	75
	Summary	82
	A World Cup Soccer Game They would Never See	83
	Human factors Analysis using HFACS	86
	Summary	90
	The Volcano Special	91
	Human Factors Analysis using HFACS	94
	Summary	97
	Conclusion	98
5	Exposing the Face of Human Error	99
	Quantifying Proficiency within the Fleet	106
	Crew Resource Management Training: Success or Failure	111
	The Redheaded Stepchild of Aviation	116
	Conclusion	121
6	Beyond Gut Feelings…	122
	Validity of a Framework	123
	Factors Affecting Validity	124
	Reliability	124
	Comprehensiveness	132
	Diagnosticity	138
	Usability	145
	Conclusion	147
7	But What About…?	149
References		*157*
Index		*163*

List of Figures

Figure 1.1	The first fatal aviation accident	2
Figure 1.2	Overall and fatal commercial air carrier accidents worldwide 1961–99	3
Figure 1.3	Accident trends for U.S. general and military aviation	4
Figure 1.4	U.S. Naval aviation accident rate and intervention strategies across calendar years 1950 to 2000	5
Figure 1.5	Original straight carrier flight deck and improved angled carrier flight deck	6
Figure 1.6	Monetary costs of accidents in the U.S. Navy/Marine Corps from fiscal year 1996 to 2000	8
Figure 1.7	Number of commercial jet accidents, accident rates, and traffic growth – past, present, and future	9
Figure 1.8	Rate of Naval aviation accidents associated with human error versus those attributable solely to mechanical or environmental factors	11
Figure 1.9	The engineering investigation and prevention process	13
Figure 1.10	Human error process loop	17
Figure 2.1	Basic model of information processing	21
Figure 2.2	Decision-making model	22
Figure 2.3	A taxonomic framework for assessing aircrew error	24
Figure 2.4	The SHEL model	27
Figure 2.5	Model of accident causation. Successful completion of the task (top); Unsuccessful completion of the task (bottom)	29
Figure 2.6	Peterson's motivation, reward, and satisfaction model	31
Figure 2.7	Epidemiological model of accident causation	33
Figure 2.8	Social factors affecting aircrew error	35
Figure 2.9	The domino theory of accident causation	38
Figure 2.10	The four "P's" of flight deck operations	41
Figure 3.1	Components of a productive system	46
Figure 3.2	The "Swiss cheese" model of accident causation	47
Figure 3.3	Categories of unsafe acts committed by aircrews	51
Figure 3.4	Categories of preconditions of unsafe acts	56

Figure 3.5	Categories of unsafe supervision	63
Figure 3.6	Organizational factors influencing accidents	66
Figure 3.7	The Human Factors Analysis and Classification System (HFACS)	71
Figure 4.1	DC-8 with engine number 1 inoperable (marked with an "X") veers left due to asymmetrical thrust from number 4 engine	74
Figure 4.2	Steps required to classify causal factors using HFACS	76
Figure 4.3	Summary of the uncontrolled collision with terrain of a DC-8 at Kansas City International Airport	83
Figure 4.4	Aircraft descent profile and ground track during the accident approach	85
Figure 4.5	Sleepiness and performance as a function of time of day	88
Figure 4.6	Summary of the controlled flight into terrain of the Learjet one mile short of Dulles International Airport	91
Figure 4.7	Planned tour route of SAT flights	92
Figure 4.8	Designated, planned, and actual flight path of SAT Flight 22	93
Figure 4.9	Summary of the in-flight collision with Mount Haleakala, Maui, Hawaii	98
Figure 5.1	Percentage and rate of U.S. Navy/Marine Corps Class A accidents associated with at least one violation as defined within HFACS	101
Figure 5.2	The percentage of U.S. Navy/Marine Corps Class A accidents associated with at least one violation as defined within HFACS. The mean percentages of Class A accidents for the U.S. Navy/Marine Corps, U.S. Army, and U.S. Air Force are plotted with dashed lines	103
Figure 5.3	The percentage of U.S. Navy/Marine Corps Class A accidents associated with at least one violation in the years before and after the intervention strategy was implemented	105
Figure 5.4	Percentage of accidents associated with skill-based errors. The linear trend is plotted as a dashed line	107
Figure 5.5	Percentage of U.S. military TACAIR and helicopter accidents occurring between FY 1991 and 2000 that were associated with skill-based errors	109

Figure 5.6	Percentage of accidents associated with decision errors. The linear trend is plotted as a dashed line	111
Figure 5.7	Percentage of accidents associated with crew resource management failures. The linear trend is plotted as a dashed line	113
Figure 5.8	Percentage of U.S. scheduled air carrier accidents associated with crew resource management failures. The linear trends for the U.S. Navy/Marine Corps and scheduled air carrier accidents are plotted as dashed lines	114
Figure 5.9	Percentage of fatal GA accidents associated with each unsafe act	118
Figure 5.10	Percentage of nonfatal GA accidents associated with each unsafe act	119
Figure 5.11	Percentage of fatal and nonfatal GA accidents associated with each unsafe act	120
Figure 6.1	Types of validity with those relevant to error taxonomies highlighted	123
Figure 6.2	Factors affecting the validity of an error-classification system	125
Figure 6.3	The process of testing and improving the reliability of an error classification system	126
Figure 6.4	The Taxonomy of Unsafe Operations	127
Figure 6.5	Modifications made to the *Taxonomy of Unsafe Operations*. Boxes outlined in dashes represent category changes. Categories deleted are indicated with an "X"	129
Figure 6.6	Additional modifications made to the *Taxonomy of Unsafe Operations*. Boxes outlined in dashes represent category changes. Categories deleted are indicated with an "X"	131
Figure 6.7	Percentage of accidents associated with perceptual errors across military and civilian aviation (1990–98)	143
Figure 6.8	Percentage of accidents associated with skill-based errors across military and civilian aviation (1990–98)	144
Figure 6.9	HFACS as modified by the Canadian Forces (CF-HFACS)	147

List of Tables

Table 2.1	Accident causation within the management system	40
Table 3.1	Selected examples of unsafe acts of operators	52
Table 3.2	Selected examples of preconditions of unsafe acts	58
Table 3.3	Selected examples of unsafe supervision	64
Table 3.4	Selected examples of organizational influences	69
Table 5.1	The number of accidents annually for U.S. commercial, military, and general aviation	116
Table 6.1	Reliability of the HFACS framework using military accident data	128
Table 6.2	The person or organization involved with a given causal factor	135
Table 6.3	What was done or not done by the individual or organization identified in Table 6.2	136
Table 6.4	Why the "what" from Table 6.3 was committed	136
Table 6.5	CFIT and non-CFIT accidents associated with at least one instance of a particular causal category	140
Table 6.6	CFIT accidents occurring in clear versus visually impoverished conditions	141

Acknowledgements

We never would have been able to write this book without the support and understanding of our loving wives. We are forever grateful for their encouragement and understanding throughout this entire endeavor. We greatly appreciate the hardships that both of them have had to endure. Not only did they have to deal with managing the home fronts and children when we went off on our "writing trips" to get away from all the distractions of our daily lives, but they also had to put up with our moodiness and despair on days when it appeared that we would never complete this book.

We would also like to thank CAPT James Fraser, COL Roger Daugherty, CAPT John Schmidt and Rear Admiral "Skip" Dirren for championing HFACS within the U.S. Navy/Marine Corps. In addition, we are indebted to Wing Commander Narinder Taneja of the Indian Air Force for his tireless efforts analyzing accident data to validate HFACS while serving as a visiting scholar at the University of Illinois. Our gratitude also goes out to both Cristy Detwiler and Karen Ayers for their assistance in editing and formatting this book.

Preface

As aircraft have become more reliable, humans have played a progressively more important causal role in aviation accidents. Consequently, a growing number of aviation organizations are tasking their safety personnel with developing accident investigation and other safety programs to address the highly complex and often nebulous issue of human error. Unfortunately, many of today's aviation safety personnel have little formal education in human factors or aviation psychology. Rather, most are professional pilots with general engineering or other technical backgrounds. Thus, many safety professionals are ill-equipped to perform these new duties and, to their dismay, soon discover that an "off-the-shelf" or standard approach for investigating and preventing human error in aviation does not exist. This is not surprising, given that human error is a topic that researchers and academicians in the fields of human factors and psychology have been grappling with for decades.

Indeed, recent years have seen a proliferation of human error frameworks and accident investigation schemes to the point where there now appears to be as many human error models as there are people interested in the topic (Senders and Moray, 1991). Even worse, most error models and frameworks tend to be either too "academic" or abstract for practitioners to understand or are too simple and "theoretically void" to get at the underlying causes of human error in aviation operations.

Having been left without adequate guidance to circumnavigate the veritable potpourri of human error frameworks available, many safety professionals have resorted to developing accident investigation and error-management programs based on intuition or "pop psychology" concepts, rather than on theory and empirical data. The result has been accident analysis and prevention programs that, on the surface, produce a great deal of activity (e.g., incident reporting, safety seminars and "error awareness" training), but in reality only peck around the edges of the true underlying causes of human error. Demonstrable improvements in safety are therefore hardly ever realized.

The purpose of the present book is to remedy this situation by presenting a comprehensive, user-friendly framework to assist practitioners in effectively investigating and analyzing human error in aviation. Coined the Human Factors Analysis and Classification System (HFACS), its framework is based on James Reason's (1990) well-known "Swiss cheese" model of accident causation. In essence, HFACS bridges the gap between theory and practice in a way that helps improve both the quantity and quality of information gathered in aviation accidents and incidents.

The HFACS framework was originally developed for, and subsequently adopted by, the U.S. Navy/Marine Corps as an accident investigation and data analysis tool. The U.S. Army, Air Force, and Coast Guard, as well as other military and civilian aviation organizations around the world are also currently using HFACS to supplement their preexisting accident investigation systems. In addition, HFACS has been taught to literally thousands of students and safety professionals through workshops and courses offered at professional meetings and universities. Indeed, HFACS is now relatively well known within many sectors of aviation and an increasing number of organizations worldwide are interested in exploring its usage. Consequently, we currently receive numerous requests for more information about the system on what often seems to be a daily basis.

To date, however, no single document containing all the information on the development and application of HFACS exists. Most of our previous work on this topic has been published in either technical reports, scientific journals or conference proceedings. Furthermore, given the development of HFACS has been an evolving process, our early publications and presentations contain much older, less complete versions of the system. Yet given the popularity and accessibility of the World Wide Web, many of these older versions are currently being circulated via documents and presentations that are available and downloadable over the Internet. As a result, some organizations are using older versions of HFACS and are not benefiting from the use of the latest and greatly improved version. Our goals in writing this book, therefore, are to integrate our various writings in this area and to expand upon them in a way not suitable for technical journals or other scientific publications. This book, therefore, will serve as a common resource for all who are interested in obtaining the most up-to-date and comprehensive description of the HFACS framework.

We have written this book primarily for practitioners (not necessarily academicians) in the field of aviation safety. Therefore, we intentionally describe human error and HFACS from an applied perspective. In doing so, our hope is that practitioners will find in this book the necessary ingredients to effectively investigate and analyze the role of human error in aviation accidents and incidents. Perhaps then, definitive improvements in aviation safety will be more readily forthcoming.

Scope of the Book

To set the stage for our discussion of HFACS, Chapter 1 provides an overview of the historical role that human error has played in aviation accidents. This chapter also examines the possible systemic reasons for the limited effectiveness of many accident prevention programs and highlights the need for the development of a comprehensive framework of human error.

Toward these ends, the prominent human error perspectives commonly discussed in the literature are presented in Chapter 2, serving as a foundation for the development of HFACS. The strengths and weaknesses of each perspective are discussed with an eye toward a unifying theory of human error that incorporates the best aspects of each.

One of the most influential unifying theories, James Reason's "Swiss cheese" model of accident causation, is presented in Chapter 3. With Reason's model as its theoretical basis, the HFACS framework is then laid out in detail to describe the latent and active failures or "holes in the cheese" as postulated by Reason.

Simply describing HFACS however, is not enough. After all "the proof of the pudding is in the eating". Therefore, a better way to gain an appreciation of how HFACS can be applied to aviation accident analysis is to demonstrate its utility using a series of case studies. With this in mind, Chapter 4 presents several examples of how HFACS can be applied to explain the human causal factors associated with actual aviation accidents.

Moving beyond the realm of accident investigation, Chapter 5 illustrates how HFACS can be used to perform comprehensive human factors analyses of existing accident databases. Examples will also be provided of how the results of such analyses have helped to identify key human factors problems within Naval aviation, so that successful interventions could be developed and implemented.

Still, how is one to know whether HFACS will have utility in an operational setting? One obvious way is simply to implement it and see if it works. However, in today's world, most organizations cannot absorb the cost in both time and money to wait and see if HFACS proves useful. Clearly, a better approach would be to use some sort of objective criteria for evaluating the framework. Chapter 6, therefore, describes the set of design criteria and the validation process used to ensure that HFACS would have utility as an accident investigation and data analysis tool.

As the final chapter, aptly named "But What About...?", Chapter 7 addresses some of the common questions and concerns that people often have about HFACS. While we would like to think that the preceding chapters adequately speak to these issues, we have chosen to meet them head-on in this chapter in order to help readers better determine the appropriateness of HFACS for their organization.

Disclaimer

The views expressed in this book are our own. They do not necessarily reflect those of the Federal Aviation Administration or the U.S. Department of Transportation. Nor do they necessarily reflect those of the U.S. Navy, Department of Defense or any other branch of the Federal Government. We have made an earnest attempt to provide proper citation to the work of others, but we do apologize if we have failed to provide appropriate credit to anyone for their efforts or ideas.

1 *Errare Humanum Est –* To Err is Human

On September 17th ... at 4:46 pm, the aeroplane was taken from the shed, moved to the upper end of the field and set on the starting track. Mr. Wright and Lieutenant Selfridge took their places in the machine, and it started at 5:14, circling the field to the left as usual. It had been in the air four minutes and 18 seconds, had circled the field 4½ times and had just crossed the aeroplane shed at the lower end of the field when I heard a report then saw a section of the propeller blade flutter to the ground. I judged the machine at the time was at a height of about 150 feet. It appeared to glide down for perhaps 75 feet, advancing in the meantime about 200 feet. At this point it seemed to me to stop, turn so as to head up the field towards the hospital, rock like a ship in rough water, then drop straight to the ground the remaining 75 feet...

The pieces of propeller blade [were] picked up at a point 200 feet west of where the airplane struck. It was 2½ feet long, was a part of the right propeller, and from the marks on it had apparently come in contact with the upper guywire running to the rear rudder. ... [The propeller] struck [the guywire] hard enough to pull it out of its socket and at the same time to break the propeller. The rear rudder then fell to the side and the air striking this from beneath, as the machine started to glide down, gave an upward tendency to the rear of the machine, which increased until the equilibrium was entirely lost. Then the aeroplane pitched forward and fell straight down, the left wings striking before the right. It landed on the front end of the skids, and they, as well as the front rudder was crushed.

Lieutenant Selfridge ... died at 8:10 that evening of a fracture of the skull over the eye, which was undoubtedly caused by his head striking one of the wooden supports or possibly one of the wires. ... Mr. Wright was found to have two or three ribs broken, a cut over the eye, also on the lip, and the left thigh broken between the hip and the knee (1st Lieutenant Frank P. Lalm, 1908).

Note, this pioneer of aviation safety was actually Frank P. Lahm, not Lalm as identified in this letter to the Chief of the Army Signal Corps.

What began as an unofficial orientation flight at Fort Meyer, Virginia in the summer of 1908, ended in tragedy, as have many flights since. Sadly, the annals of aviation history are littered with accidents and tragic losses such as this (Figure 1.1).

Since the late 1950s, however, the drive to reduce the accident rate has yielded unprecedented levels of safety. In fact, today it is likely safer to fly in a commercial airliner than to drive a car or walk across a busy New York City street. Still, it is interesting that while historians can recount in detail the strides that the aviation industry has made over the last half century, one fundamental question remains generally unanswered: *"Why do aircraft crash?"*

Figure 1.1 The first fatal aviation accident
Source: arlingtoncemetary.com

The answer may not be as straightforward as you think. For example, in the early years of aviation it could reasonably be said that the aircraft itself was responsible for the majority of aircraft accidents. That is, early aircraft were intrinsically unforgiving and, relative to their counterparts today, mechanically unsafe. However, the modern era of aviation has witnessed an ironic reversal of sorts. It now appears to some that the aircrew themselves are more deadly than the aircraft they fly (Mason, 1993; cited in Murray, 1997). Indeed, estimates in the literature indicate that somewhere between 70 and 80 percent of all aviation accidents can be attributed, at least in part, to human error (Shappell and Wiegmann, 1996).

So, maybe we *can* answer the larger question of why aircraft crash, if only we could define what really constitutes that 70 to 80 percent of human error referred to in the literature. But, even if we did know, could we ever really hope to do anything about it? After all, *errare humanum est* – to err is human (Plutarch, c.100 AD). So, isn't it unreasonable to expect error-free human performance? Maybe ... but, perhaps a lesson in how far aviation safety has come since its inauspicious beginnings nearly a century ago will provide us with some clues about where we need to go next.

Aviation Safety Trends

Most aviation accident statistics cited in the literature today begin with data collected in the late 1950s and early 1960s. Representative of this type of data are the two graphs presented in Figure 1.2. In the top graph, the number of commercial aviation accidents that have occurred worldwide since 1961 are plotted annually against the number of departures. When the data are

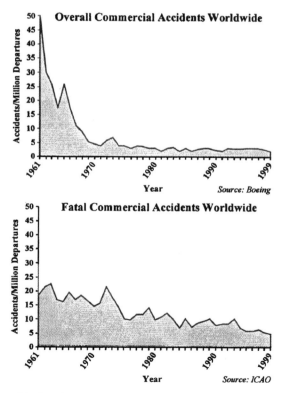

Figure 1.2 Overall (top) and fatal (bottom) commercial air carrier accidents worldwide 1961–99

depicted in this manner, a sharp decline in the accident rate since the early 1960s becomes readily apparent. In fact, the number of commercial accidents has decreased to a point where today, fewer than two accidents occur worldwide for every one million departures (Boeing, 2000; Flight Safety Foundation [FSF], 1997). What's more, this trend is generally the same (albeit not as dramatic), whether you consider the overall number of commercial aviation accidents, or just those associated with fatalities. In either case, it can reasonably be said that commercial aviation safety has steadily improved over the last 40 years. Indeed, aviation has become one of the safest forms of transportation, leading the National Transportation Safety Board to proclaim in 1990 that a passenger boarding a U.S. carrier then had over a 99.99 percent chance of surviving the flight (NTSB, 1994a).

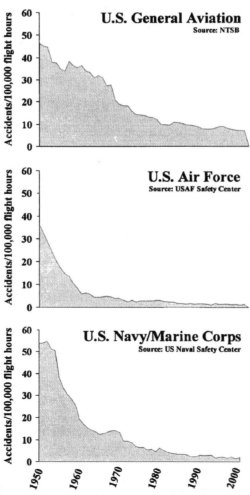

Figure 1.3 Accident trends for U.S. general and military aviation

Improvements in aviation safety, however, are not unique to commercial aviation. General aviation accident rates have also plummeted over the last several decades (Figure 1.3, top). A similar trend can also be seen when accident data from the U.S. Navy/Marine Corps (middle) and U.S. Air Force (bottom) are plotted across years. Indeed, the accident rates among these diverse types of flying operations have dropped impressively since the late 1950s and early 1960s, indicating that *all* aspects of aviation have benefited from advances aimed at making the skies safer.

So, what can we attribute these improvements in aviation safety to over the last half-century? Given the rather dramatic changes evident in the accident record, it is doubtful that any single intervention was responsible for the decline in the accident rate. Rather, it is likely the result of a variety of factors, such as advancements in technology, equipment, operating procedures, and training practices (Nagel, 1988; Yacavone, 1993).

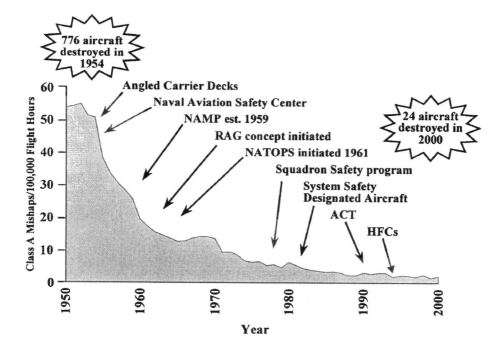

Figure 1.4 U.S. Naval aviation accident rate and intervention strategies across calendar years 1950 to 2000
Source: *U.S. Naval Safety Center*

To give you a better feel for how various interventions have improved aviation safety, let us consider several of these initiatives within the context of Naval aviation. In Figure 1.4, a number of technical innovations and standardization programs introduced into the U.S. Navy/Marine Corps over the last several decades have been superimposed on the annual accident rate. Arguably, these efforts were not solely responsible for the decline observed in the accident rate. After all, nowhere does this figure address improvements in aircraft design and the introduction of new aircraft in the fleet. Still, there is little question among Naval experts that these interventions played a significant role in the level of safety currently enjoyed by the U.S. Navy/Marine Corps.

CV-6 USS ENTERPRISE
Straight Carrier Deck

CVN-73 USS GEORGE WASHINGTON
Angled Carrier Deck

Figure 1.5 Original straight carrier flight deck (top) and improved angled carrier flight deck (bottom)
Source: *U.S. Navy*

Consider, for example, the development of the angled carrier deck aboard Naval aircraft carriers in the early to mid-1950s. Many Naval history buffs may recall that early aircraft carrier flight decks were single straight runways, which created a number of safety problems – especially when one aircraft was trying to take off from the bow of the ship while another was unexpectedly aborting a landing on the stern (Figure 1.5, top). Not surprising, aircraft would occasionally collide! To remedy this safety hazard, the angled carrier deck was developed, which allowed aircraft to take off from the bow of the ship in a different direction from those landing on the stern, avoiding any potential conflict in their flight paths; a very wise intervention indeed (Figure 1.5, bottom).

Another major factor affecting safety in the U.S. Navy/Marine Corps was the establishment of the Naval Aviation Safety Center (now known as the Naval Safety Center) in the mid-1950s. On the surface, this might not seem to be particularly revolutionary given today's standards. However, for the first time, a major command in the U.S. Navy was assigned the sole responsibility and authority for monitoring and regulating safety issues in the fleet. This single act elevated aviation safety to the highest levels of the U.S. Navy/Marine Corps, as the command reported directly to the Chief of Naval Operations.

Still, other safety programs have helped improve Naval aviation safety as well. For example, in the early 1960s, the replacement air group concept was created, requiring pilots to receive specialized training in advanced aircraft before flying them in the fleet. While it may sound intuitive to some that pilots should gain some tactical experience in their aircraft before flying them in combat or other operations, this was not always the case. As recently as WWII, pilots would receive basic flight training and then transition to the fleet, entering the operational arena with very little time in their combat aircraft.

More recently, the establishment of formal squadron safety programs, the development of aircrew coordination training, and the implementation of a periodic human factors review of fleet aviators have all contributed significantly to Naval aviation safety by identifying problems and hazards before they resulted in accidents. Undeniably, safety initiatives such as these have saved countless lives in the U.S. Navy/Marine Corps and have elevated Naval aviation safety to unprecedented levels.

Beyond saving lives, the military, like any other business, is often driven by the so-called "bean counters." Yet, even the bean counters have to be smiling when you consider the cost savings realized as a result of improvements in aviation safety. Consider that until recently the U.S. Navy/Marine Corps flew an average of 2 million flight hours per year (today it's closer to 1.5 million flight hours per year). If the rate of major accidents today were still at levels observed in 1950, over 800 aircraft would have been lost in 2000 alone! Needless to say, the U.S. Navy/Marine Corps would be quickly out of the aviation business altogether, if that were the case. Thankfully, improvements in all forms of aviation safety, including Naval aviation, have remedied this trend.

Some Reasons for Concern

Even though the overall accident rate in civil and military aviation is indeed excellent, certain aspects of the data are "unsettling" (Nagel, 1988, p. 264). As can be seen from the graphs presented earlier, improvements in aviation safety have slowed substantially during the last few decades. This plateau has led some to conclude that further reductions in the accident rate are improbable, if not impossible. In other words, we have reached a point at which accidents may simply be the "cost of doing business." However, if we accept this philosophy we must also be prepared to accept the consequences. For example, on the military side of aviation, the financial cost of aircraft accidents is growing astronomically. As illustrated in Figure 1.6, the amount of money and resources lost due to U.S. Naval aviation accidents is enormous, even though these accidents occur much less frequently than other types. Indeed, the loss incurred from aviation accidents cost the U.S. Navy/Marine Corps some 3.3 billion in the last five years alone – more than five times that seen with all other accidents combined. Obviously, if the mishap rate were allowed to continue at its current level, either taxes would have to go up to buy more airplanes (not a politically popular option), or the military would have to operate with fewer and fewer aircraft (not a strategically savvy move either).

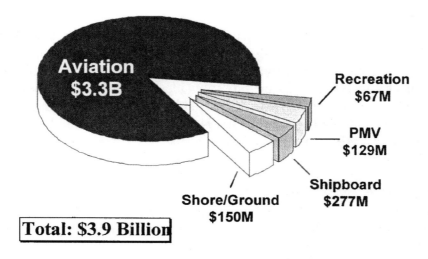

Figure 1.6 Monetary costs of accidents in the U.S. Navy/Marine Corps from fiscal year 1996 to 2000
Source: *Fraser (2002)*

There may be reason for concern within commercial aviation as well. Consider, for example, that worldwide air traffic is expected to increase significantly over the next several years as the industry continues to grow (FSF, 1997). Now let's assume for the moment that the current commercial accident rate is already "as good as it's going to get." Naturally, if you increase the number of flights while maintaining the same accident rate, the overall frequency of accidents will inevitably increase as well. To illustrate this point, the current commercial jet accident rate, expected traffic growth, and frequency of accidents have been plotted together in Figure 1.7. Sadly, if these estimates remain unchanged, there may be as many as 50 major airline accidents occurring worldwide per year during the first decade of the new millennium. This equates to nearly one accident a week!

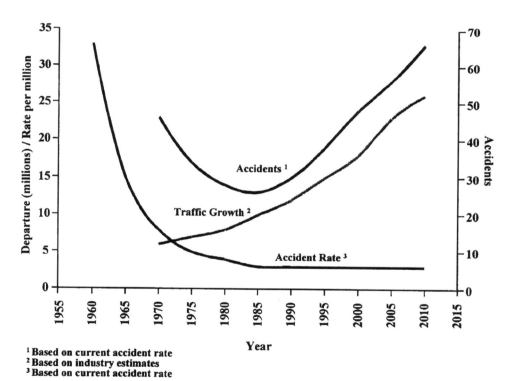

[1] Based on current accident rate
[2] Based on industry estimates
[3] Based on current accident rate

Figure 1.7 Number of commercial jet accidents, accident rates, and traffic growth – past, present, and future
Source: Flight Safety Foundation (1997)

Given the intense media coverage that major airline accidents often receive, combined with the rapid dissemination of information worldwide, there is little doubt that the traveling public will be made well aware of these accidents in the most explicit detail, even if they do occur half-way around the world. As such, the airline industry will likely suffer as public confidence erodes and a general mistrust permeates the aviation industry. Simply trotting out the industry "talking heads" and releasing statements such as "the accident rate has not changed" or that "we are as safe as we have ever been" will likely have little or no effect on public confidence, nor will it likely appease the flying public's demand for the safest form of transportation possible.

One alternative may be to post the NTSB safety statistic cited earlier on the bulkhead of each airplane. Can you imagine reading the following placard as you board the airplane with the rest of your family for that fun-filled trip to Disneyland?

> **Welcome aboard Doug and Scott's airline. You have a 99.99% chance of surviving this flight.**

Not a particularly comforting thought, is it? Well ... then again, public relations were never our strong suit. Beside, the NTSB statistic cited earlier only refers to survival. There are no guarantees that you will not be involved in an accident or maimed – only that you will likely survive the ordeal.

But seriously, "accident-prevention steps must be taken now to stop the accident rate from exceeding its current level, and even greater effort must be taken to further reduce the current accident rate" (FSF, 1997). After all, even if the industry was willing to accept the monetary cost of accidents, the loss of lives alone makes further reductions a necessity, not a commodity to be traded. Still, the days of sweeping reductions and sharp drops in the accident rate due to a few innovations or interventions have been over for nearly 30 years. Any change will likely be measured as a reduction in only a few accidents a year – and the cost of those interventions will be the result of millions of dollars worth of research and investigation. Therefore, with limited budgets and the stakes so high, accident prevention measures must target the primary cause of accidents, which in most cases, is the human (ICAO, 1993).

Human Error and Aviation Accidents

Recall, that roughly 70 to 80 percent of all aviation accidents are attributable, at least in part, to some form of human error. Notably, however, as the accident rate has declined over the last half century, reductions in human error-related accidents have not kept pace with the reduction of accidents related to mechanical and environmental factors (NTSB, 1990; Nagel, 1988; O'Hare et al., 1994; Shappell and Wiegmann, 1996; Yacavone, 1993). In fact, humans have played a progressively more important causal role in both civilian and military aviation accidents as aircraft equipment has become more reliable (Nagel, 1988). For example, our previous analysis of Naval aviation mishap data (Shappell and Wiegmann, 1996), revealed, that in 1977, the number of Naval aviation accidents attributed solely to mechanical and environmental factors was nearly equal to those attributable, at least in part, to human error (Figure 1.8). Yet, by 1992, the number of solely mechanical accidents had been virtually eliminated, while the number of human-error related accidents had been reduced by only 50 percent. We have even argued that the reduction in accidents attributable to human error was not as much a function of interventions aimed at aircrew, as it was improvements made to the aircraft. After all, it is well known that the opportunity for human error will go up considerably when a mechanical failure occurs. Not surprising then, as aircraft have become more reliable, accidents due to human error would naturally decline as well.

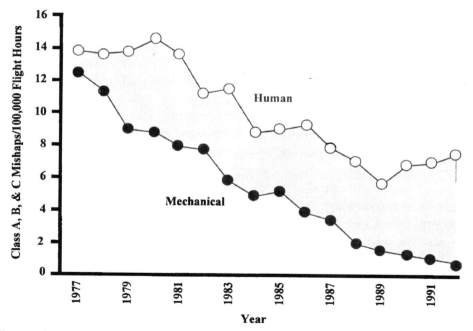

Figure 1.8 Rate of Naval aviation accidents associated with human error versus those attributable solely to mechanical or environmental factors

So it would appear that many of the interventions aimed at reducing the occurrence or consequence of human error have not been as effective as those directed at mechanical failures. Undeniably, there are many reasons for this disparity – some more obvious than others. Regardless of the reasons however, they can all be best understood within the context of the accident investigation and prevention process. Therefore, let us consider in more detail the differences in the accident investigation and intervention process from both the engineering and human factors side of the house. Although both processes normally occur simultaneously, each will be considered separately to illustrate their inherent differences.

Engineering Aspects of an Investigation

Although much less frequent today than in years past, mechanical failures occasionally do occur in flight, and in worst-case scenarios may even lead to an incident or accident as illustrated in Figure 1.9. Typically, an investigation will then take place involving a team of air-safety investigators and technical support personnel charged with sifting through the wreckage to uncover hidden clues as to the accident's cause. Collectively, this investigative team possesses a wide range of experience, including specialized knowledge of aircraft systems, aerodynamics, and other aerospace engineering topics. In addition, these highly trained accident sleuths often have access to an assortment of sophisticated technology and analytical techniques such as metallurgical tests, electron microscopy, and advanced computer modeling capabilities, all designed to enrich the investigative process.

Armed with a blend of science and sophisticated instrumentation that would make even James Bond green with envy, it is no surprise that most, if not all, mechanical failures that result in accidents are often revealed during the engineering investigation. To illustrate this point, let us suppose for a moment that the structural integrity of an aircraft is compromised by fatigue fractures along a wing spar or among a series of bolts or rivets. These fractures, when viewed with an electron microscope, have unique patterns that can be easily identified by experienced engineers and metallurgists, leaving little doubt as to the origin of the failure. In much the same way, the presence of a system malfunction can be uncovered by a detailed examination of the electrical wiring of the aircraft, including the breakage pattern of light bulb filaments within the instrument panel. For example, if a particular system warning light was illuminated at the time of impact (presumably indicating that the system was inoperative) there is a distinctive stretch to the white-hot filament within the bulb. Combined with other

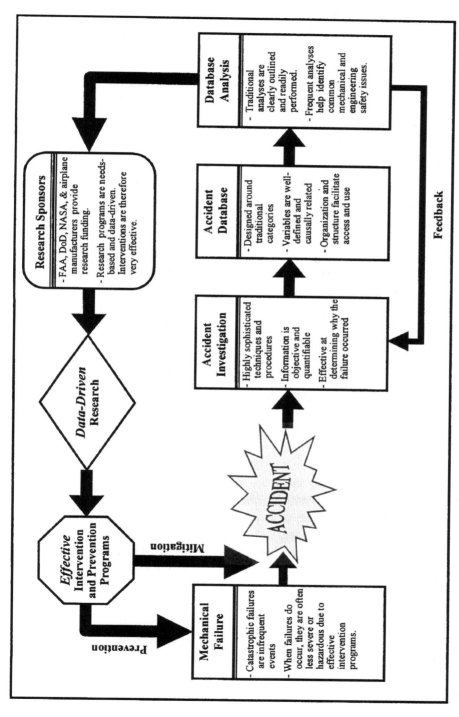

Figure 1.9 The engineering investigation and prevention process

supporting evidence such as frayed electrical wires, it can then be determined if a system failure contributed in a significant way to the accident.

Regardless of the methods employed, what makes evidence gathered in engineering investigations so indisputable is that the techniques and analyses involved are grounded in the physical sciences. This fact alone allows investigators to move beyond simply identifying and cataloging what part of the aircraft failed, to the larger question of why the failure occurred in the first place. As a result, data gathered in engineering investigations have yielded revolutionary design changes and have contributed significantly to the evolution of today's modern aircraft.

The collection of accident data alone, however, would be of little use if a repository/database did not exist to house it. Typically then, data from engineering investigations are entered into accident databases maintained by safety organizations like the National Transportation Safety Board (NTSB) in Washington, DC. Such databases are generally highly structured and well defined, being organized around traditional aircraft categories such as airframes, powerplants, and component systems. As a result, the data are easily accessible, allowing periodic analyses to be performed so that major causal trends or common problems can be identified across accidents.

The results from these analyses in turn provide feedback to investigators, which improves their investigative methods and techniques while providing guidance on where to look during future investigations. For example, if analysts at the NTSB were to find that a particular engine had a history of fatigue related failures, then this information could be distributed to investigators in the field for use during their next investigation. In effect, the accident database provides a rich source of clues when investigating future accidents.

In addition, information from the database analyses provides a valuable resource for researchers within the FAA, NASA, DoD and airplane manufacturers whose mission is to develop safer and more efficient aircraft. Ultimately, these needs-based, data-driven programs produce effective intervention strategies that either prevent mechanical failures from occurring or mitigate their consequences when they do. What's more, given that these interventions are "data-driven," their effectiveness can be objectively monitored and evaluated, so that they can be modified or reinforced to improve safety. The result of this engineering investigative and prevention process has been a dramatic reduction in the rate of accidents due to mechanical or systems failures.

Human Factors Aspects of an Investigation

In contrast to the engineering investigation just described, consider the occurrence of an aircrew error that results in an accident (Figure 1.10). As with mechanical failures, an investigation soon takes place to determine the nature and cause of these errors. However, unlike the engineering investigation that involved numerous technical experts, the human performance investigation typically involves only a single individual, who may or may not be trained in human factors. In fact, even at the world's premier safety organizations there may be only a handful of human factors professionals on the staff. Truth be told, if you were to knock on the door of the NTSB or any of the U.S. military safety centers today and ask them to send out their human factors experts, only a few people would exit the building. Now, ask them to send out their engineering experts. It would look like a fire drill, as practically the whole building empties! Perhaps this is a bit of an exaggeration, but the point is that most human performance investigators are often a "one person show", with little assistance or support in the field or elsewhere.

What makes matters worse is that unlike the tangible and quantifiable evidence surrounding mechanical failures, the evidence and causes of human error are generally qualitative and elusive. Even the analytical techniques used within the human factors investigation are generally less refined and sophisticated than those employed to analyze mechanical and engineering concerns. Consider, for example, the difference between fatigue in a bolt and a fatigued pilot. Unlike metal fatigue that can be readily identified using well-established technology and electron microscopy, pilot fatigue is difficult to observe directly, much less quantify. Instead, it must be inferred from a variety of factors such as the time an accident occurred and the pilot's 72-hour history, which includes, among other things, when he/she went to bed and how long they slept. In addition, other issues such as work tempo, experience, and flight duration may also come into play, all of which make any determination of pilot fatigue an inexact science at best. So, while engineers have little difficulty agreeing upon fatigue in a bolt, it remains virtually impossible to get a group of accident investigators to agree on the presence of fatigue in a pilot, even if all of the necessary information is available.

Like pilot fatigue, the identification of other human factors causal to an accident is easier said than done. As a result, human factors investigations have traditionally focused on "what" caused the accident, rather than "why" it occurred. Indeed, many human causal factors in accident reports are not "really causes on which safety recommendations can be made, but rather merely brief descriptions of the accident" or error (ICAO, 1993, p. 32).

Statements like the pilot "failed to maintain adequate clearance from the terrain" provide little insight into possible interventions. In effect, the only safety recommendations that could be derived from such a statement would be to either make a rubber airplane or make rubber ground – neither of which make much sense outside the confines of children's cartoons!

Still, investigators identify human causal factors, and as with the engineering side of the investigation, the information gathered during the human performance investigation is entered into an accident database. However, unlike their engineering counterparts, databases that house human error data are often poorly organized and lack any consistent or meaningful structure. This should come as no surprise when you consider that "information management" technicians who possess expertise in archiving data but have little familiarity with human factors, design most accident databases. As a result, these data warehouses are quite effective in preserving the data (much like mummification preserves the body), but they have proven woefully inadequate for data retrieval and analysis. In fact, as the ardent researcher unwraps the proverbial database mummy, there is often considerable disappointment as he soon discovers that what's inside bears little resemblance to traditional human factors. That is to say, there is generally no theoretical or functional relationship between the variables, as they are often few in number and ill defined.

Given the dearth of human factors data and the inherent problems associated with most databases, when aviation accident data are examined for human error trends, the result is typically less than convincing. Accordingly, many safety professionals have labeled the entire contents of the database as "garbage," a view not appreciated by those doing the investigations. Still, even with its shortcoming, analysts and academicians continue to wrestle with the data and are resolved to making something out of their contents. Unfortunately, many of these analyses simply focus on more reliable contextual information such as time of day, weather conditions, and geographic location of the accident or demographic data surrounding accidents, such as pilot gender, age, and flight time. In fact, few studies have attempted to examine the underlying human causes of accidents. Even those have generally been limited to a small subset of accidents that often only relate to the researchers particular area of interest. Rarely, if ever, has there been a comprehensive and systematic analysis of the *entire* database to discover the major human factors issues related to flight safety.

Results from analyses of accident data have therefore provided little feedback to help investigators improve their investigative methods and techniques. The information is also of limited use to airlines and government agencies in determining the types of research or safety programs to sponsor. Not surprising then, many human factors safety programs tend to be

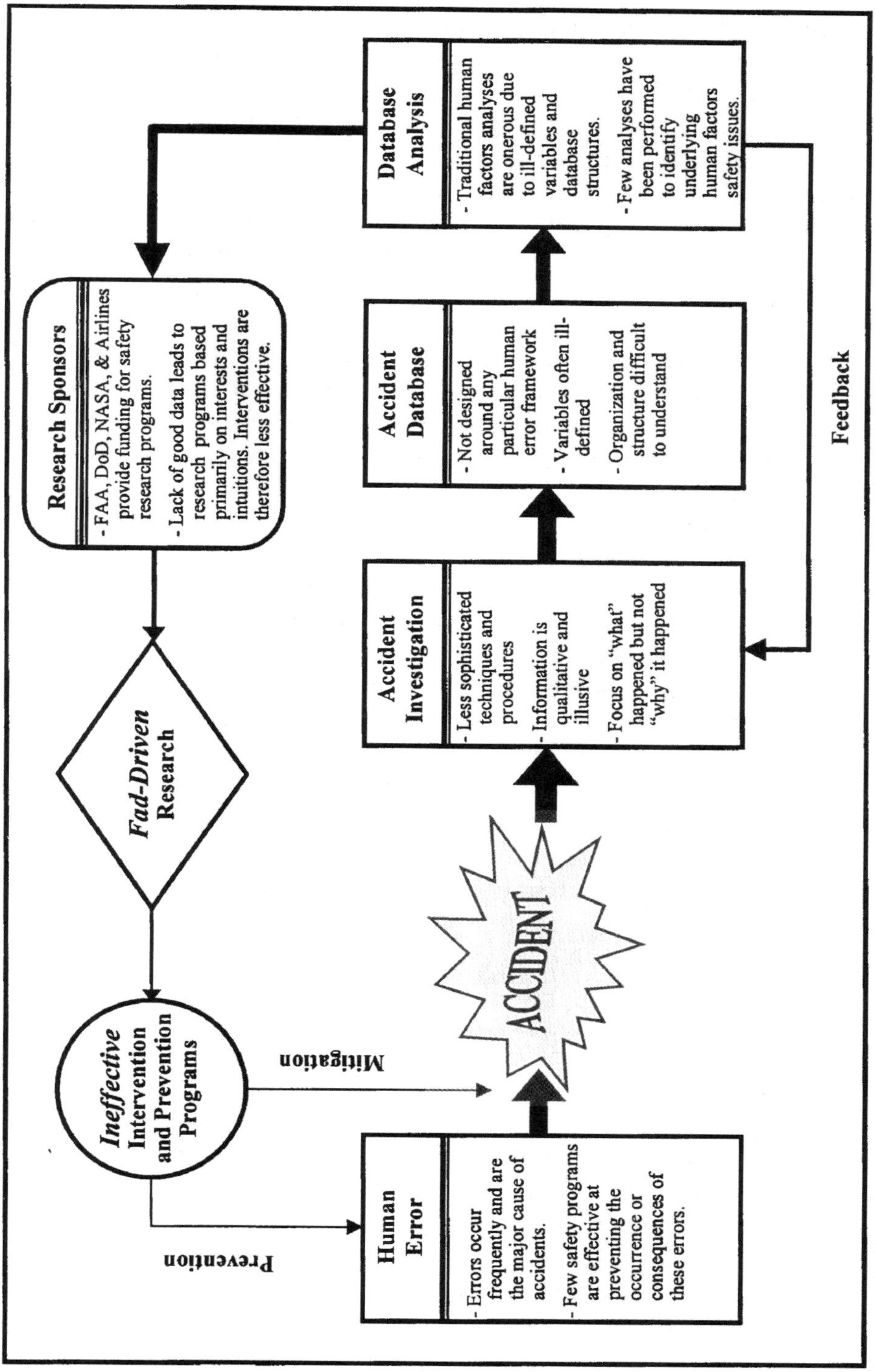

Figure 1.10 Human error process loop

intuitively- or fad-driven, rather than the data-driven programs initiated within the engineering side of the house. That is to say, interventions aimed at human factors are typically derived by well-meaning, "expert" opinion or group discussions about what many "believe" are the major safety issues. In truth however, many decisions about safety programs are based on statements like, "I've flown the line, and never crashed from being fatigued, so fatigue cannot be a big problem," or "the last accident was due to CRM problems, therefore we need to spend more money on improving CRM."

Curiously, most would admit that this opinion-based process would not work on the engineering side. Imagine an engineer standing up in a meeting and emphatically stating that he or she has a "gut feeling" about the airworthiness of a particular aircraft. Such a statement not based on data, would clearly result in more than just a few odd looks from co-workers if not outright ridicule. Nevertheless, such is often the status quo on the human factors side and many don't think twice about it!

Given that most human factors safety programs are not data-driven, it only stands to reason that they have produced intervention strategies that are only marginally effective at reducing the occurrence and consequences of human error. Furthermore, unlike the engineering side in which single interventions can often produce great strides in improving the structural integrity and reliability of mechanical systems, human factors interventions are often constrained by the limited improvements that can be achieved in the performance capabilities of humans. What's more, the lack of consistent human factors accident data has prohibited the objective evaluation of most interventions so that they might be revamped or reinforced to improve safety. As a result, the overall rate of human-error related accidents has remained high and constant over the last several years (Shappell and Wiegmann, 1996).

Conclusion

The current aviation safety system was built on issues that confronted aviation 50 years ago, when the aircraft was, in effect, the "weakest link." Today, however, accidents attributable to catastrophic failures of the aircraft are very infrequent. If the aviation industry is ever to realize a reduction in the aviation accident rate, the human causes of accidents need to be more effectively addressed.

However, simply replacing all of the engineers and other technical experts with those versed in human factors is not the solution. That would be like "throwing the baby out with the bath water" and would likely result in an increase in accidents attributable to mechanical and engineering factors.

Instead, the human factors aspects of aircraft accident investigations need to be enhanced. Nevertheless, one does not necessarily need a doctorate in human factors to perform a legitimate human performance investigation. Current air-safety investigators could effectively assume these responsibilities. This is not to say, however, that simply having a brain by default makes an engineer or a pilot a human factors expert. Just because we all eat, doesn't make us all experts in nutrition. Air-safety investigators need to be provided with a better understanding of human factors issues and analytical techniques.

Increasing the amount of money and resources spent on human factors research and safety programs is not necessarily the answer to all of our safety problems either. After all, a great deal of resources and efforts are currently being expended and simply increasing these efforts would likely not make them more effective. To paraphrase Albert Einstein, "the definition of insanity is doing something over and over again and expecting different results." Instead, the solution may be to redirect safety programs so that they address important human factors issues.

Regardless of the mechanism, safety efforts cannot be systematically refocused until a thorough understanding of the nature of human factors in aviation accidents is realized. Such an understanding can only be derived from a comprehensive analysis of existing accident databases. What is required to achieve these objectives is a general human error framework around which new investigative methods can be designed and existing post-accident databases restructured. Such a framework would also serve as a foundation for the development and tracking of intervention strategies, so that they can be modified or reinforced to improve safety. The question remains, as to whether such a human error framework exists – a topic we turn to in the next chapter.

2 Human Error Perspectives

Recent years have witnessed a proliferation of human error frameworks to a point where today there appears to be as many human error models and taxonomies as there are people interested in the topic (Senders and Moray, 1991). What remains to be answered, however, is whether any of these frameworks can actually be used to conduct a comprehensive human error analysis of aviation accident data and/or provide a structure around which new human factors investigative techniques can be designed. After all, if an adequate "off-the-shelf" approach for addressing human error already exists, it would eliminate the need to develop yet another error framework. In other words, why reinvent the wheel if you don't have to? This is the very question that we have wrestled with within our own organizations.

So how do you identify the right error framework for your purposes? Perhaps the best way is to do what we did and systematically examine the approaches others have taken to address human error (Wiegmann and Shappell, 2001a). Only then can you accurately determine which frameworks, if any, are suitable to meet your needs.

At first glance, such a task can be daunting, particularly if one tries to survey each, and every one of the error frameworks that exist. However, what we have found is that when these different methods are sorted based upon the underlying assumptions made about the nature and causes of human error, a smaller, more manageable, collection of error systems will emerge. Using this approach, our previous forays into the human error literature have revealed six major human error perspectives, all of which have distinct advantages and disadvantages (Wiegmann et al., 2000; Wiegmann and Shappell, 2001a). In no particular order, they include the cognitive, ergonomic, behavioral, aeromedical, psychosocial, and organizational perspectives.

In the next several pages we will explore each of these human error perspectives, focusing on selected frameworks that characterize each approach as well as their strengths and weaknesses. Then, after reviewing each perspective, we will once again return to the question of whether any of the existing frameworks provide a suitable foundation for conducting a comprehensive analysis of human error associated with aviation accidents and incidents.

The Cognitive Perspective

Let us begin by first examining one of the more popular approaches to human error analysis – the cognitive perspective. The principle feature of this approach is the assumption that the pilot's mind can be conceptualized as essentially an information processing system. Much like a modern computer, the cognitive viewpoint assumes that once information from the environment makes contact with one of the senses (e.g., vision, touch, smell, etc.), it progresses through a series of stages or mental operations, culminating in a response.

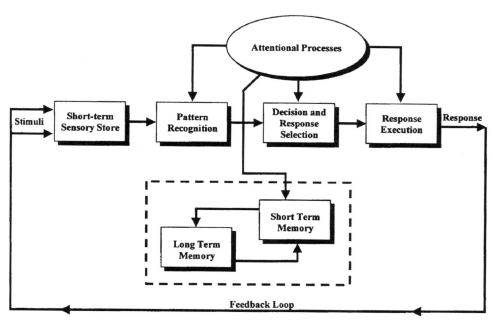

Figure 2.1 Basic model of information processing
Source: Adapted from Wickens and Flach (1988)

The four-stage model of information processing described by Wickens and Flach (1988) is but one example of this view (Figure 2.1). In their model, stimuli from the environment (e.g., photons of light or sound waves) are converted into neural impulses and stored temporarily in a short-term sensory store (e.g., iconic or echoic memory). Provided sufficient attention is devoted to the stimulus, information from the short-term sensory store is then compared with previous patterns held in long-term memory to create a mental representation of the current state of the world. From there, individuals must decide if the information they glean requires a response or can simply be ignored until something significant occurs. But, let us assume for the moment that something important has happened, like an engine fire, and that a specific action is necessary to avert disaster. In this eventuality, information would then be passed to the response execution stage where the selection of appropriate motor programs would occur, enabling the pilot to activate the appropriate engine fire extinguishers. Still, the process doesn't stop there as the response is monitored via a sensory feedback loop, which in this case would ensure that the fire was put out, and if not, would stimulate the system to make the necessary modifications and adjustments until the situation was resolved.

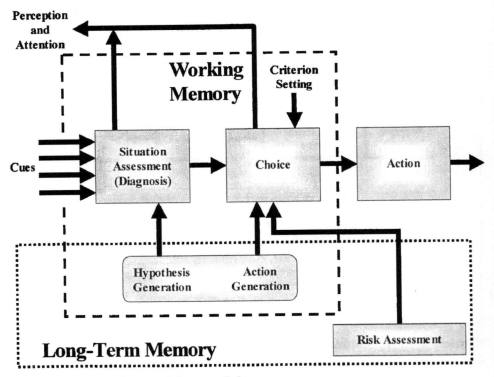

Figure 2.2 Decision-making model
Source: Adapted from Wickens and Flach (1988)

Using this four-stage model of information processing, Wickens and Flach (1988) proposed the general model of decision-making presented in Figure 2.2. In their model, an individual will sample a variety of cues in their environment to assess a given situation. These cues are then compared against a knowledge base contained within long-term memory so that an accurate diagnosis of the situation can take place. Then, given that a problem has been identified, choices have to be made regarding what action, or actions, should be taken. This process requires an evaluation of possible actions and utilizes risk assessment and criterion setting to ensure that an appropriate response will be employed. What's more, at any point in this decision-making process, individuals can seek out additional information (indicated by the lines to *perception and attention*) to improve situational assessment or enhance their response.

Unfortunately, errors can arise at many points during this process. For example, cues can be absent or barely perceptible resulting in a poor or inaccurate assessment of the situation. Then again, individuals may correctly assess their current state of affairs, but choose the wrong solution or take unnecessary risks, resulting in failure. In fact, everything can be processed correctly and the right decision made, yet the pilot may not possess the skills necessary to avert disaster. Regardless of where the failure occurs, by capitalizing on our understanding of human information processing capabilities, decision-making models such as the one proposed by Wickens and Flach provide insight into why errors are committed, and why accidents happen.

Using this same approach, Rasmussen (1982) developed a detailed taxonomic algorithm for classifying information processing failures. This algorithm, as employed within the context of aviation (e.g., O'Hare et al., 1994; Wiegmann and Shappell, 1997; Zotov, 1997), uses a six-step sequence to diagnose the underlying cognitive failure(s) responsible for an error (Figure 2.3). As described by O'Hare et al. in 1994, the algorithm includes stimulus detection, system diagnosis, goal setting, strategy selection, procedure adoption, and action stages, all of which can either fail independently or in conjunction with one another to cause an error.

As one might expect, there is significant overlap between elements of Rasmussen's taxonomy and the four-stage model of information processing described earlier. For instance, Rasmussen's information processing errors correspond closely with the input of cues and short-term sensory storage of Wickens and Flach. Likewise, Rasmussen's diagnostic errors fit nicely with the pattern recognition stage, while goal, strategy and procedure errors are closely matched with decision-making and response selection. Finally, elements of Wickens and Flach's response execution stage are captured within Rasmussen's final category of action errors.

Given the step-by-step, logical approach of cognitive models like the two presented above, this perspective remains popular among academicians and aviation psychologists for analyzing human error in complex systems. However, their appeal to those who actually do accident investigations is largely because they attempt to go beyond simply classifying "what" the aircrew did wrong (e.g., the pilot failed to lower the landing gear or the aircraft was flown into the terrain) to addressing the underlying causes of human error (e.g., the failure of attention, memory or specific types of decision errors). As a result, these cognitive models allow seemingly unrelated errors to be analyzed based on fundamental cognitive failures and scientific principles.

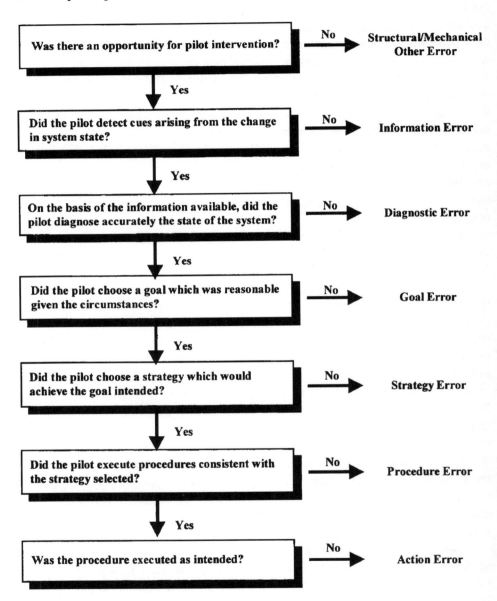

Figure 2.3 A taxonomic framework for assessing aircrew error
Source: *Adapted from O'Hare et al. (1994)*

Wiegmann and Shappell (1997), for example, used three cognitive models, including the four-stage model of information processing and the modified Rasmussen model to analyze over 4,500 pilot-causal factors associated with nearly 2,000 U.S. Naval aviation accidents. Although the models differed slightly in the types of errors that they captured, all three generally converged on the same conclusion. That is, judgment errors (e.g., decision making, goal setting and strategy selection errors) were associated more often with major accidents, while procedural and response execution errors were more likely to lead to minor accidents.

These findings make intuitive sense if you consider them within the context of automobile accidents. For instance, if your timing is off a bit on the brake or your driving skill leaves something to be desired, our findings suggest that the odds are you are more likely to be involved in a minor fender-bender. On the other hand, if you elect to "run" a stoplight or drive at excessive speeds through a school zone, our findings would indicate that you are more likely to be involved in a major accident, or even worse, you may kill someone or yourself! But we were not the first ones to see this. In fact, findings similar to ours were found with other military (Diehl, 1992) and civilian aviation accidents (O'Hare et al., 1994; Jensen and Benel, 1977) using the cognitive approach. In the end, studies such as these have helped dispel the widely held belief that the only difference between a major accident and so-called "fender-bender" is little more than luck and timing. Indeed, it now appears to be much more.

In theory, a better understanding of the types of cognitive failures that produce errors would in turn, allow for the identification and development of effective intervention and mitigation strategies. According to the cognitive perspective, these interventions would target the pilots' information processing capability. However, unlike computers that can be improved by simply upgrading the hardware, the information processing hardware of the human (i.e., the brain) is generally fixed inside the head. Therefore, in order to improve performance, cognitive psychologists typically attempt to capitalize on the manner in which pilots process information. For example, examining how expert pilots solve problems or distribute their attention in the cockpit can help scientists develop better methods for training novice aircrew. Another way of improving information processing is through the standardization of procedures and the use of checklists. These methods often facilitate information processing by reducing mental workload and task demands during normal operations and emergencies, thereby reducing the potential for errors and accidents.

Nevertheless, as popular and useful as cognitive models are, they are not without their limitations where accident investigation is concerned. For instance, many cognitive theories are quite academic and difficult to translate into the applied world of error analysis and accident investigation. As a result, the application of these theoretical approaches often remains nebulous and requires analysts and investigators to rely as much on speculation and intuition as they do on objective methods. What's more, cognitive models typically do not address contextual or task-related factors such as equipment design or environmental conditions like temperature, noise, and vibration. Nor do they consider conditions like fatigue, illness, and motivational factors, all of which impact pilot decision-making and information processing.

Perhaps more important however, supervisory and other organizational factors that often impact performance are also overlooked by traditional cognitive models. Consequently, those that espouse the cognitive approach have been accused of encouraging an extreme, almost single-minded view that focuses solely on the operator (aircrew) as the "cause" of the error. This sort of single-mindedness often results in blame being unduly placed on the individual who committed the error rather than on its underlying causes which the individual may have little or no control over. Within the context of aviation, this view is sustained by those who regard pilots as the major cause of aircraft accidents or the weak link in the aviation safety chain. In effect then, pilots may be viewed as more dangerous than the aircraft they fly (Mason, 1993; cited in Murray, 1997). Clearly, such extreme views are detrimental to aviation safety in general, and may ultimately limit the advancement of the cognitive approach.

The Ergonomic Perspective

Now let us turn to the ergonomic or "systems perspective." According to this approach, the human is rarely, if ever, the *sole* cause of an error or accident. Rather, human performance involves a complex interaction of several factors including "the inseparable tie between individuals, their tools and machines, and their general work environment" (Heinrich, et al., 1980, p. 51).

Perhaps the most well known of the systems perspectives is the SHEL model proposed by Edwards (1988), which describes four basic components necessary for successful man–machine integration and system design (Figure 2.4). SHEL, in this case, is an acronym representing the four components of the model, the first of which is software, represented by the letter "S". However, unlike the computer software we are all familiar with today, here software represents the rules and regulations that govern how a system operates. The "H," on the other hand, refers to the hardware associated with a given system, such as the equipment, material, and other physical assets. The "E" refers to the environment and was created to account for the physical working conditions that we as humans (liveware – symbolized by the letter L) are faced with.

Edwards, recognizing that the four components of the SHEL model do not act in isolation, highlighted the interactions between components (indicated by the links in Figure 2.4). He felt that it was at the boundaries of these interfaces that many problems or mismatches occur. Within aviation for example, the focus has historically been on the liveware–hardware (better known as human–machine) interface, yielding significant improvements in cockpit layout and other so-called "knobs and dials" issues. In fact, the match between the human and the equipment within a given environment is viewed as so crucial to aircraft development today that human factors principles are often considered throughout the design process.

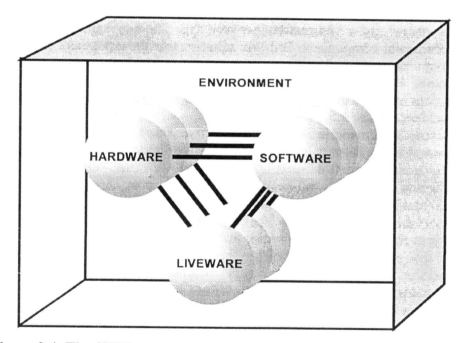

Figure 2.4 The SHEL model
Source: Adapted from Edwards (1988)

However, even the two-dimensional interfaces between components do not sufficiently describe the SHEL model, as multi-dimensional models are more typical of normal day-to-day operations within a given system (represented by the multiple spheres in Figure 2.4). For example, with the development of datalink communications in aviation, the so-called liveware–hardware–liveware interface has been of great concern. In fact, before datalink is instituted, engineers and scientists will have to demonstrate that the proposed modifications to the liveware (air traffic controller) – hardware (datalink technology) – liveware (pilot) interface will enhance pilot–controller communications, or at a minimum produce no decrement in performance (Prinzo, 2001). Unfortunately, these multi-dimensional interactions are often hidden from the operator, producing opaque systems that, if not designed properly, can detract from the monitoring and diagnosing of system problems, thereby producing accidents (Reason, 1990).

As popular as the SHEL model is, it is not the only ergonomics show in town. One alternative is the model of accident causation proposed by Firenze in 1971. His model is based on the premise that humans will make decisions based upon information they have acquired. Obviously, the better the information, the better the decision, and vice-versa. These decisions will allow the individual to take certain risks to complete a task (Figure 2.5, top).

Like the SHEL model, Firenze's model predicts that system failures occur when there is a mismatch between the human, machine, and/or environmental components. But this assumes that the equipment (machine) functions properly and that the environment is conducive to a successful outcome. Problems arise when stressors such as anxiety, fatigue, and hazardous attitudes distort or impede the decision making process and lead to an accident (Figure 2.5, bottom).

Clearly, improving information can prevent some accidents, but this may over-emphasize failures associated with the human, equipment, and/or environment – a point not lost on Firenze who felt that, "the probability of eliminating all failures where man interacts with machines is practically zero." Making matters worse, the environment often exacerbates whatever stressors an individual may be feeling at a given point in time. So if your goal is to reduce accidents, Firenze, and those that espouse his views, would argue that efforts must focus on the system as a whole, not just the human component.

A close examination of the systems approach reveals some clear advantages over the cognitive failure models described earlier. For example, the systems perspective considers a variety of contextual and task-related factors that effect operator performance, including equipment design. In doing so, system models discourage analysts and investigators from focusing solely on the operator as the source or cause of errors. As a result, greater varieties of error prevention methods are available, including the possibility of designing systems that are more "error-tolerant."

System approaches also have an intuitive appeal, particularly to those not formally trained in aviation psychology or human factors. In particular, approaches such as Edwards' SHEL model are very easy to comprehend, are relatively complete from an engineering point of view, and are generally well known across disciplines. In fact, in 1993, the International Civil Aviation

Organization (the body governing aviation worldwide) recommended the use of the SHEL model as a framework for analyzing human factors during aviation accident investigations. Other organizations like the U.S. Air Force and Air Line Pilots Association have based portions of their investigative framework on this system as well.

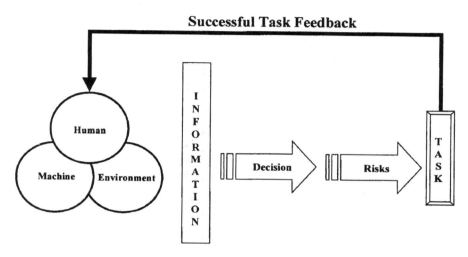

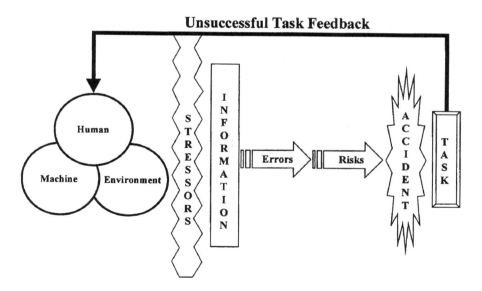

Figure 2.5 Model of accident causation. Successful completion of the task (top); Unsuccessful completion of the task (bottom)
Source: *Adapted from Firenze (1971)*

Nevertheless, even with their apparent popularity, the generality afforded by system models often comes at the cost of specificity. For instance, most system models lack any real sophistication when it comes to analyzing the human component of the system. Since system models focus on the interaction among components, emphasis is placed almost exclusively on the design aspects of the man–machine interface (e.g., the design of knobs, dials and displays), as well as the possible mismatch between the anthropometric requirements of the task and human characteristics. The effects of cognitive, social, and organizational factors therefore receive only tacit consideration, giving the impression that these components of the system are relatively unimportant. As a result, the systems perspective tends to promulgate the notion that all errors and accidents are design-induced and can therefore be engineered out of the system – a view not universally held within the aviation safety community.

The Behavioral Perspective

The behavioral perspective deals with the topic of pilot performance and aircrew error a bit differently than either the cognitive or ergonomic approaches. Rather than emphasizing an individual's ability to process information or how one integrates into the system as a whole, behaviorists believe that performance is guided by the drive to obtain rewards and avoid unpleasant consequences or punishments (Skinner, 1974).

For example, the motivation-reward-satisfaction model proposed by Peterson in 1971, describes performance as dependent upon one's innate ability and motivation, which in turn is dependent upon a number of other factors (Figure 2.6). For instance, personnel selection plays a large role in determining whether someone has the aptitude to succeed; yet, without adequate training, performance will likely suffer. Likewise, motivation is critical to optimal performance regardless of where that motivation comes from – whether from the job, peers, unions, or internally derived.

But motivation and ability alone cannot fully explain how people behave. Indeed, the cornerstone of Peterson's model is the extent to which individuals feel satisfied about their performance, which in turn is largely dependent on the rewards that they receive within an organization. Even one's sense of accomplishment and pride in a job well done can serve as a reward and thereby effect satisfaction. Ultimately, it is this feeling of satisfaction that motivates individuals to perform the same action again and again.

Human Error Perspectives 31

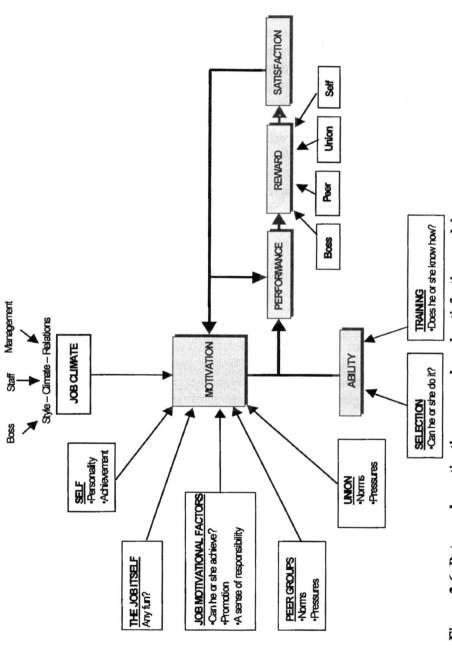

Figure 2.6 Peterson's motivation, reward, and satisfaction model
Source: Adapted from Peterson (1971)

Behavioral models like Peterson's have contributed greatly to our understanding of how factors such as motivation, rewards, and past experience affect performance and safety. For instance, when individuals lack either the motivation to perform safely, or when conditions exist that reward unsafe actions, rather than those that are safe, accidents will likely occur. Even in the aviation industry, where safety is often highlighted, there are situational factors that reinforce unsafe behavior or punish individuals who emphasize safety at the expense of other organizational concerns. What's more, safe actions seldom lead to immediate tangible rewards, but serve only to prevent the occurrence of something aversive (e.g., crashing the airplane). It is not surprising then that some pilots have been known to bend, or worse yet, break the rules. As a result, recent years have witnessed the rise of behavioral-based safety programs that seek to reward safe behavior while at the same time enforcing the rules and holding aircrew and their supervisors accountable when unsafe acts occur.

Even with its obvious benefits however, aviation safety professionals have never fully embraced the behavioral perspective. Still today, many question its applicability. This may be due in part to the fact that within the realm of aviation safety, the consequences of unsafe behavior are often fatal, and therefore it is hard to believe that someone would not be motivated to perform at their best. As Fuller (1997) has noted, "Perhaps we don't ask about motivation for air safety for the same reasons we don't ask about the motivation for breathing" (p. 175). Beyond that, it is hard to imagine how actions like misreading a flight management system (FMS) or forgetting to lower the landing gear can be linked to motivational factors. Still, there are unsafe acts that are obviously connected to motivation and should not be ignored during accident investigations. As a result, some human factors professionals and researchers, such as Reason (1990), have begun to distinguish between unsafe acts that are motivation-driven (i.e., violations) and those that are truly cognitive in nature (i.e., errors). Such a distinction is indeed important when it comes to developing interventions for reducing unsafe acts and improving safety.

The Aeromedical Perspective

Based largely upon the traditional medical model, the aeromedical perspective has been championed by those who feel that errors are merely the symptoms of an underlying mental or physiological condition such as illness or fatigue. The belief is that these so-called "pathogens" exist insidiously within the aircrew until they are triggered by environmental conditions or situations that promote their manifestation as symptoms (errors). In fact, some theorists believe that physiology affects virtually all aspects of safe behavior (Reinhart, 1996), and that the concept of being "medically airworthy" goes hand-in-hand with aviation safety and performance.

Using traditional medical research methods, some safety experts have taken an epidemiological approach to analyzing accidents. The most common of which are those done by consumer product safety groups. One of the early epidemiological models of accident causation proposed by Suchman in 1961 is presented in Figure 2.7. Suchman's model is analogous to those used in medicine today to study the host, agent, and environmental factors that cause diseases. When applying the model, the investigator seeks an explanation for the occurrence of an accident within the host (accident victim), the agent (injury or damage deliverer), and environmental factors (physical, social and psychological characteristics of a particular accident setting).

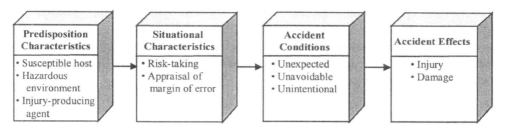

Figure 2.7 Epidemiological model of accident causation
Source: Adapted from Suchman (1961)

Above all else, the aeromedical approach highlights the crucial role that the physiological state of the pilot (i.e., the host) plays in safe performance and flight operations (Lauber, 1996). Although this may seem painfully obvious to those in the aerospace medicine community, others have not always taken the aeromedical perspective seriously. For example, while military pilots have long been taught about the adverse effects of hypoxia, decompression sickness, spatial disorientation, and other physiological factors by flight surgeons and aerospace physiologists, training in flight physiology within the civilian sector has typically been minimized. As a result, civilian pilots often have little respect for the significance of these factors within aviation (Reinhart, 1996).

One aeromedical factor that has received considerable attention over the years in both military and civilian aviation is fatigue. As knowledge of the physiological underpinnings of circadian rhythms and jet lag has developed, an awareness of the adverse impact that fatigue has on aircrew performance has grown. This mounting appreciation was strengthened by the NTSB (1994b) ruling that identified fatigue as a causal, rather than contributory, factor in an airline accident – one of the first rulings of its kind in the history of the Board. Without a doubt, the aeromedical community has taken the lead in shaping both the military's and industry's view of fatigue and has helped form policies on such contentious issues as work scheduling, shift-rotations, and crew-rest requirements.

As with the other perspectives, the aeromedical approach is not without its critics. As mentioned above, some view pilot physiology and factors that influence it as relatively unimportant in the big picture of flight safety. This may be due to the fact that some pilots find it difficult to understand how adverse physiological states such as decompression sickness, trapped gases, and gravity-induced loss of consciousness (G-LOC) impact pilot performance in modern commercial and general aviation. Or even more unclear is how fatigued, self-medicated, or disoriented a pilot has to be before he or she commits an error that fatally jeopardizes the safety of flight. In short, it is not difficult to imagine how the presence of such factors may "contribute" to an error, but determining whether these factors "caused" an error or accident is another matter entirely. Although this "cause-effect" problem may seem trivial to some, to others in the aviation industry it weighs heavily on how resources and manpower are allocated to improve safety within their organizations.

The Psychosocial Perspective

The psychosocial perspective, unlike the others reviewed thus far, takes a more humanistic approach to behavior. Those that champion this approach view flight operations as a social endeavor that involves interactions among a variety of individuals, including pilots, air-traffic controllers, dispatchers, ground crew, maintenance personnel, and flight attendants. Incredibly, this cast of players from seemingly disparate organizations works closely together to ensure the level of safety we all enjoy in aviation today. Even the private pilot is seldom, if ever, entirely alone in the air or on the ground as air traffic control is only a button push away.

These delicate, yet complex, interactions are at the center of the psychosocial perspective. Indeed, many aviation psychologists and safety professionals alike believe that pilot performance is directly influenced by the nature or quality of the interactions among group members (Helmreich and Foushee, 1993). These interactions in turn are influenced not only by the operating environment but also by the personalities and attitudes of individuals within each group. Given the inherent diversity and the sheer number of individuals involved day-to-day, one can only marvel at the precision and level of safety that modern aviation enjoys. According to this perspective, it is only when the delicate balance between group dynamics and interpersonal communications breaks down that errors and accidents occur (Figure 2.8).

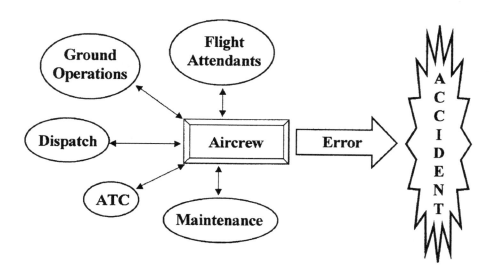

Figure 2.8 Social factors affecting aircrew error
Source: *Adapted from Helmreich and Foushee (1993)*

Historically, psychosocial models have been overlooked by those in the aviation industry (Kayten, 1993). In fact, it has only been in the past decade that aviation psychologists and accident investigators have truly begun to study the interpersonal aspects of human performance when examining aircrew errors. One such study involved an industry-wide analysis of aviation accidents and found that over 70 percent of all accidents resulted from aircrew coordination and communication problems (Lautman and Gallimore, 1987). However, this finding is not unique to commercial aviation. Aircrew coordination failures have been recognized as a major cause of military aviation accidents as well (Wiegmann and Shappell, 1999; Yacavone, 1993). As a result of these and other studies in the literature, many conventional engineering psychologists are now reaching beyond traditional design issues of the human–machine interface and beginning to address the exceedingly complex issues of human interpersonal relationships. Even those who promote the cognitive approach have begun to consider the possible impact that social factors have on processes such as decision making (Orasanu, 1993).

As a direct result of the large number of accidents involving simple communication failures, more and more intervention strategies are being aimed at improving cockpit communications, including crew resource management (CRM) training. A staple within modern military and commercial aviation, CRM training involves educating and training aircrew to use techniques that enable individuals to communicate problems more effectively, divide task responsibilities during high workload situations, and resolve conflicts in the cockpit. In fact, one of the early aims of CRM training was to challenge and change pilots' traditional attitudes about differences in authority between the captain and the other aircrew (e.g., the co-pilot or first officer), an area that has been shown to impede communication and cause accidents (Wiegmann and Shappell, 1999). There is little debate that improvements in aircrew coordination and communication have reduced errors in the cockpit (Kern, 2001) and improved aviation safety.

However, the psychosocial perspective hasn't always enjoyed the popularity within the aviation industry that it does today. This may be because many of the early approaches focused largely on personality variables rather than on crew coordination and communication issues that most contemporary approaches do. One of these early models included the concept of accident proneness, arguing that some individuals were simply predisposed toward making errors and causing accidents (Haddon et al., 1964). Today, however, the idea that accidents are inevitable among certain individuals is difficult for most theorists to accept. As a result, such fatalistic views have quickly fallen out of favor.

But even more radical are those models that were based upon traditional psychoanalytic (Freudian) views of human behavior, which suggest that errors and accidents are caused by an individual's unconscious desire to harm others or to gratify unfulfilled sexual wishes. The following is an excerpt from Brenner (1964) illustrating this perspective:

> ...while driving her husband's car, [a woman], stopped so suddenly that the car behind her crumpled one of the rear fenders of the car she was in. The analysis of this mishap revealed a complicated set of unconscious motives. Apparently, three different, though related ones were present. For one thing, the [woman] was unconsciously angry at her husband because of the way he mistreated her. As she put it, he was always shoving her around. Smashing up his car was an unconscious expression of this anger, which she was unable to display openly and directly against him.

For another thing, she felt very guilty as a result of what she unconsciously wanted to do to her husband in her rage at him and damaging his car was an excellent way to get him to punish her. As soon as the accident happened, she knew she was 'in for it.' For a third thing, the [woman] had strong sexual desires which her husband was unable to satisfy and which she herself had strongly repressed. These unconscious, sexual wishes were symbolically gratified by having a man 'bang into [her] tail,' as she put it (p. 295).

At the time, Brenner (1964) concluded that psychoanalytic explanations of accident causation were of "sufficient interest, influence, and plausibility to justify their scientific evaluation." However, in reality, such views have almost always been far outside the mainstream. Indeed, psychoanalytical models of accident causation and others like them were eventually rejected on both empirical and theoretical grounds.

In fact, it can be argued that even current theories are at risk of suffering the same fate if more is not done to firm up the underlying psychosocial mechanisms that presumably lead to errors in the cockpit. With few exceptions (e.g., Helmreich and Foushee, 1993; Orasanu, 1993), little work has been done to empirically test predictions derived from psychosocial models of human error. Instead, most supporters of the psychosocial approach often reference the accident statistics cited earlier (e.g., Lautman and Gallimore, 1987; Wiegmann and Shappell, 1999; Yacavone, 1993) as confirmation of their perspective. However, these accident data are the very same data that were used to formulate such models and therefore, cannot logically be used again in reverse as supportive evidence.

This lack of clarity is effected even more by the all-encompassing definition of CRM currently used in the industry, which describes CRM as the "effective use of all available resources [by the cockpit crew], including human resources, hardware, and information" (FAA, 1997, p. 2). As an anonymous reviewer once noted – given this "...broad definition, one might conclude that the only human error mishap [not caused by] CRM failures would be the deliberate crashing of the aircraft by a depressed or otherwise disturbed crew member." Indeed, what once appeared to be a useful concept has been expanded to a point where it may have lost some of its value.

The Organizational Perspective

Organizational approaches to understanding human error have been utilized in a variety of industrial settings for many years. However, it is only recently that the aviation community has embraced this point of view. This may be due to the fact that during the early days of aviation, emphasis was placed solely on the aircraft and those that flew them. Only now, are safety practitioners realizing the complex nature of accident/incident causation and the role organizations (not just aircrew and aircraft) play in the genesis and management of human error. In fact, it is the emphasis that organizational models place on the fallible decisions of managers, supervisors, and others in the organization that sets them apart from other perspectives.

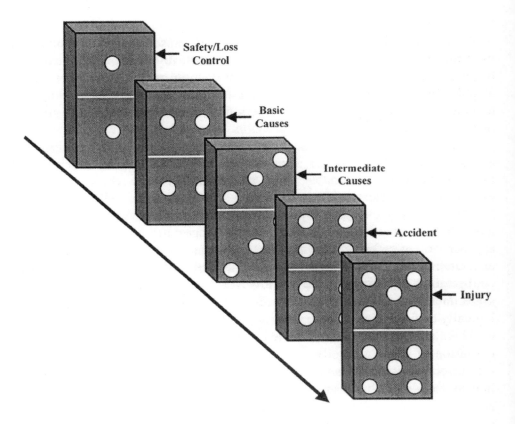

Figure 2.9 The domino theory of accident causation
Source: Adapted from Bird (1974)

Perhaps the best-known organizational model of human error is the so-called "Domino Theory" described by Bird in 1974 (Figure 2.9). Bird's theory is based in large part on the premise that, "the occurrence of an [accident] is the natural culmination of a series of events or circumstances, which invariably occur in a fixed and logical order" (Heinrich et al., 1980, p. 23). That is, much like falling dominoes, Bird and others (Adams, 1976; Weaver, 1971) have described the cascading nature of human error beginning with the failure of management to control losses (not necessarily of the monetary sort) within the organization. Exactly how management does this is often difficult to put your finger on. What we do know is that virtually all managers are tasked with identifying and assigning work within the organization, establishing performance standards, measuring performance, and making corrections where appropriate to ensure that the job gets done. If management fails at any of these tasks, basic or underlying personal (e.g., inadequate knowledge/skill, physical and mental problems, etc.) and job-related factors (e.g., inadequate work standards, abnormal usage, etc.) will begin to appear. Often referred to as origins or root causes, these basic causes often lead to what Bird referred to as immediate causes that have historically been the focus of many safety programs. Specifically, immediate causes are those unsafe acts or conditions committed by employee/operators such as the unauthorized use of equipment, misuse of safety devices, and a veritable potpourri of other unsafe operations. Ultimately, it is these immediate causes that lead to accidents and injury.

Several other organizational theorists have built upon Bird's Domino Theory, including the aforementioned Adams (1976) and Weaver (1971). For example, Adams renamed and expanded dominos one, two, and three to include elements of management structure, operational errors, and tactical errors respectively (Table 2.1). In so doing, Adams built upon Bird's original theory to more thoroughly address the relative contributions of employees, supervisors, and management to accident causation. Note, for example, that tactical errors focus primarily on employee behavior and working conditions, while operational errors are associated more with supervisory and manager behavior. Even domino one (Bird's "Loss Control") was modified to capture aspects of management structure that were not addressed by Bird. In many ways, what Adams really did was "operationalize" Bird's original ideas for use in industry – an approach still in vogue today in many settings.

Weaver's (1971) contribution, although earlier than Bird's published ideas in 1974, viewed dominoes three, four and five as symptoms of underlying operational errors, much like Adams five years later. But Weaver's aim was to expose operational error by examining not only "what caused the accident", but also "why the unsafe act was permitted and whether supervisory-management had the safety knowledge to prevent the accident" (p. 24). In other words, was management knowledgeable of the laws, codes, and standards associated with safe operations; and if so, was there confusion on the part of employees regarding the goals of the organization, the roles and responsibilities of those participating in the work setting, accountability, and the like? Questions like these probe deeper into the underlying cause of operational errors, which all three theorists (Adams, Bird and Weaver) believe are founded in management.

Table 2.1 Accident causation within the management system[1]

MANAGEMENT STRUCTURE	OPERATIONAL ERRORS		TACTICAL ERRORS		ACCIDENT/ INCIDENT	INJURY/ DAMAGE
	Manager Behavior	Supervisor Behavior	Employee Behavior (Unsafe Acts)	Work Conditions (Unsafe Conditions)		
Objectives Goals Standards Appraisals Measurements **Organizations** Chain of Command Span of Control Delegation of Authority **Operations** Equipment Scheduling Procedures Environment	Policy Goals Authority Responsibility Accountability Span of Attention Delegation	Conduct Responsibility Authority Rules Coaching Initiative Morale Operations	Poor Teamwork Using Unsafe Equipment Failure to Use Protective Gear Failure to Warn Others Using Equipment in an Unsafe Manner Horseplay	Improper Design Poor Housekeeping Improperly Guarded Improper Illumination Improper Ventilation	Injury Producing Near-miss/No Injury Property Damage Incident	To Persons To Property

[1] From 'Accident Causation and the Management System' by E.E. Adams, 1976, *Professional Safety*, October, p. 26. Copyright 1976 by American Society of Safety Engineers. Reprinted with permission.

Source: *Adapted from Adams (1976)*

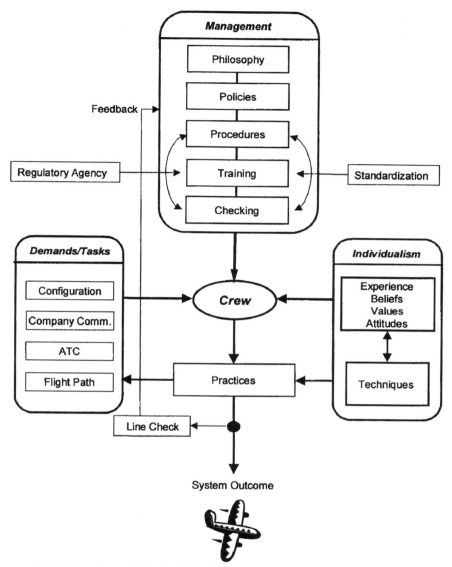

Figure 2.10 The four "P's" of flight deck operations
Source: Adapted from Degani and Wiener (1994)

A slightly different approach has been proposed by Degani and Wiener (1994) for operations on the flight deck. As illustrated in Figure 2.10, their approach focuses on the relationship between the four "P's" of flight deck operations: 1) Management's *philosophy* or broad-based view about how they will conduct business; 2) *Policies* regarding how operations are to be performed; 3) *Procedures* and/or specifications concerning how certain actions are to be executed; and 4) *Practices* of aircrew as they perform their flight-related duties. According to Degani and Weiner, all of these factors interact to enhance flight safety. However, whenever ambiguous philosophies, policies, procedures, and practices exist, or when conflicts between the four "P's" arise, safety is jeopardized and accidents can occur.

Consider, for example, the goal of most commercial airlines to have a safe and on-time departure. For obvious reasons, such a policy is critical to the success of any airline and deeply rooted within the management structure. Indeed, it is not surprising to see company slogans and advertisements promoting this philosophy. In translating philosophy into policy, many airlines have developed extensive plans to ensure that servicing of aircraft, routine maintenance, refueling, crew manning, and passenger and baggage loading all take place in a well orchestrated and lock-step fashion to ensure that the aircraft pushes back from the ramp "on-time." The procedures themselves are even more detailed as specific checks in the cockpit, sign-offs by ramp and gate personnel, and a variety of other safety and service checks are done in a very orderly and timely fashion. Ultimately however, the entire process is dependent on the men and women who perform these functions, and in doing so, put the philosophy, policies, and procedures of the organization into practice.

That being said, the entire system can break down if, for example, the philosophy of the organization drives policies that are motivated more by profit than safety (e.g., an on-time departure at all costs). Misguided corporate attitudes such as these can lead to poor or misinterpreted procedures (e.g., abbreviated cockpit checklists or the absence of a thorough aircraft walk-around) and worse yet, unsafe practices by aircrew and other support personnel. Thankfully, this is rarely, if ever, the case.

Traditional organizational theories like the Domino Theory and the Four P's have quickly gained acceptance within the aviation community and as with the other perspectives, have much to offer. The good news is that with the organizational perspective, aviation also gets the rich tradition and long established field of industrial and organizational (I/O) psychology. In fact, the methods that have proven valuable for error and accident prevention in aviation are not unlike those used in other industrial settings for controlling the quality, cost, and quantity of production (Heinrich et al., 1980). Consequently, the principles and methods developed and studied by I/O psychologists to improve worker behavior for decades (e.g., selection, training, incentives, and organizational design) should also be effective at reducing human error in aviation.

Another advantage of the organizational approach is that it views all human error as something to be managed within the context of risk. The benefits of this operational risk management approach is that it allows the importance of specific errors to be determined objectively based on the relative amount of risk they impose on safe operations. This concept has not been lost on the U.S. military, as all branches utilize risk management in some fashion within their aviation and other operations.

But even before the organizational perspective was considered within the aviation community, traditional I/O concepts had been employed for years. For example, to ensure that only skilled and safe pilots got into the cockpit, airlines and others within the aviation community employed the use of pilot selection tests. For those organizations that train their own pilots, as do most militaries around the world, these selection tests attempt to "weed out" those applicants who exhibit less than adequate mental aptitudes or psychomotor skills necessary for flying. Even organizations that hire pre-trained pilots (either from the military or general aviation sectors) often use background and flight experience as employment criteria, while others also use medical screenings and interviews to select their pilots.

Another organizational approach, as illustrated by the Degani and Wiener (1994) model, to reducing errors in the cockpit is through the establishment of *policies or rules* that regulate what pilots can and cannot do in the cockpit. Such rules may restrict the type of weather in which pilots may operate their aircraft, or may limit the number of hours pilots can spend in the cockpit, in order to avoid the possible detrimental effects of fatigue on performance. By placing only safe and proficient pilots in the cockpit and limiting aircraft operations to only safe flying conditions, organizations are able to reduce the likelihood that pilots will make mistakes and cause accidents.

Still, some have criticized that the "organizational causes" of operator errors are often several times removed, both physically and temporally, from the context in which the error is committed (e.g., the cockpit). As a result, there tends to be a great deal of difficulty linking organizational factors to operator or aircrew errors, particularly during accident investigations. Worse yet, little is known about the types of organizational variables that actually cause specific types of errors in the cockpit. Therefore, the practicality of an organizational approach for reducing or preventing operator error has been drawn into question. Furthermore, as with the other approaches described earlier, organizational models tend to focus almost exclusively on a single type of causal-factor (in this case, the fallible decisions of officials within the management hierarchy, such as line managers and supervisors) rather than the aircrew themselves. As a result, organizational models tend to foster the extreme view that "every accident, no matter how minor, is a failure of the organization" or that "...an accident is a reflection on management's ability to manage...even minor incidents are symptoms of management incompetence that may result in a major loss" (Ferry, 1988).

Conclusion

The preceding discussion of the different human-error perspectives was provided to help synthesize the different approaches or theories of human error in aviation. However, as mentioned earlier, there is no consensus within the field of aviation human factors regarding human error. Therefore, some human factors professionals may take issue, or at least partially disagree, with the way in which one or more of these perspectives and example frameworks were characterized or portrayed. Although this may provide academic fodder for those in the human factors field, that was not the intent of this chapter. Rather, the purpose was to address the question of whether any of these existing frameworks provides a foundation for conducting a comprehensive human error analysis of aviation accidents and incidents.

The answer to the above question is clear. Although each human error perspective has its own strengths, each also has inherent weaknesses. Therefore, none of the perspectives reviewed, in and of themselves, were able to address the plethora of human causal factors associated with aviation accidents. So this leads us to the next logical question, "can an approach or model be developed that captures and capitalizes on the various strengths of each approach while eliminating or reducing their limitations?" We will explore the answer to this question in the next chapter.

3 The Human Factors Analysis and Classification System (HFACS)

Several theorists and safety professionals have proposed "unifying frameworks" for integrating the diverse perspectives and models of human error described in the previous chapter (e.g., Degani and Wiener, 1994; Sanders and Shaw, 1988). While a few have enjoyed limited success, none has come close to the almost universal acceptance and praise that James Reason has received for his model of accident causation. The approach offered by Reason (1990) has literally revolutionized contemporary views of safety within aviation and throughout other industrial settings. For that reason alone, it is worth spending a few pages summarizing his perspective.

Reason's Model of Accident Causation

Elements of a Productive System

Originally developed for the nuclear power industry, Reason's approach to accident causation is based on the assumption that there are fundamental elements of all organizations that must work together harmoniously if efficient and safe operations are to occur. Taken together, these elements comprise a "productive system" as depicted in Figure 3.1.

Based on this model, the aviation industry can be viewed as a complex productive system whose "product" is the safe conduct of flight operations, regardless of whether it was for transportation, recreation, or national defense. As with any productive system, one of the key elements is the activity of front line operators (pilots, in the case of aviation) at the "pointy end" of the spear. These so-called "productive activities," in turn, require the effective integration of human and mechanical elements within the system, including among other things, effective pilot–cockpit interfaces so that safe flight operations can take place.

46 A Human Error Approach to Aviation Accident Analysis

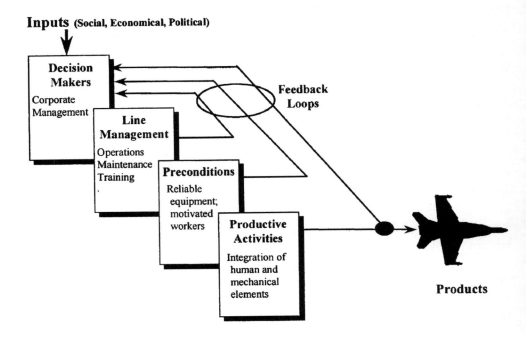

Figure 3.1 Components of a productive system
Source: *Adapted from Reason (1990)*

Before productive activities can occur, certain "preconditions" such as reliable and well-maintained equipment, and a well-trained and professional workforce, need to exist. After all, few pilots are independently wealthy and own their own airline or fleet of aircraft. Rather, they dutifully work within a highly structured organization that requires effective management and careful supervision. Furthermore, such management and supervision is needed across numerous departments within the organization, including among others, operations, maintenance, and training.

Even the best managers need guidance, personnel, and money to perform their duties effectively. This support comes from *decision-makers* who are even further up the chain-of-command, charged with setting goals and managing available resources. These same individuals have the unenviable task of balancing oft-competing goals of safety and productivity, which for airlines includes safe, on-time, cost-effective operations. Still, executive decisions are not made in a vacuum. Instead, they are typically based on social, economic, and political *inputs* coming from outside the organization, as well as *feedback* provided by managers and workers from within.

In most organizations, the system functions well. But, what about those rare occasions when the wheels do come off? Now, the system that only moments earlier appeared safe and efficient, can find itself mired in doubt and mistrust by the workforce and those that it serves. Unfortunately, this is where many safety professionals are called in to pick up the pieces.

Breakdown of a Productive System

According to Reason, accidents occur when there are breakdowns in the interactions among the components involved in the production process. These failures degrade the integrity of the system making it more vulnerable to operational hazards, and hence more susceptible to catastrophic failures. As illustrated in Figure 3.2, these failures can be depicted as "holes" within the different layers of the system; thereby transforming what was once a productive process into a failed or broken down one. Given the image of Swiss cheese that this illustration generates, the theory is often referred to as the "Swiss cheese" model of accident causation.

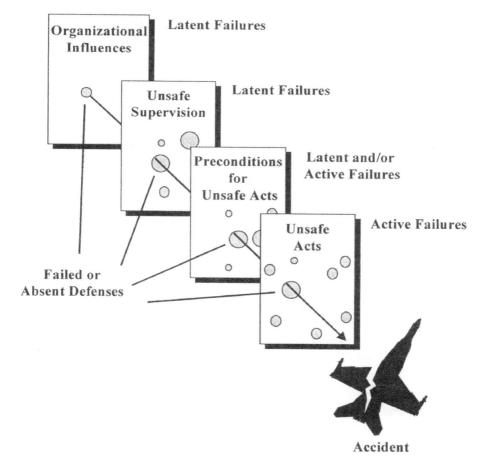

Figure 3.2 The "Swiss cheese" model of accident causation
Source: Adapted from Reason (1990)

According to the "Swiss cheese" model, accident investigators must analyze all facets and levels of the system to understand fully the causes of an accident. For example, working backwards in time from the accident, the first level to be examined would be the *unsafe acts* of operators that have ultimately led to the accident. More commonly referred to in aviation as aircrew/pilot error, this level is where most accident investigations typically focus their efforts and consequently, where most causal factors are uncovered. After all, it is these *active failures*, or actions of the aircrew, that can be directly linked to the event. For instance, failing to lower the landing gear, or worse yet, improperly scanning the aircraft's instruments while flying in instrument meteorological conditions (IMC), may yield relatively immediate, and potentially grave, consequences. Represented as failed defenses or "holes" in the cheese, these active failures are typically the last unsafe acts committed by aircrew.

However, what makes the "Swiss cheese" model particularly useful in accident investigation is that it forces investigators to address latent failures within the causal sequence of events as well. As their name suggests, latent failures, unlike their active counterparts, may lie dormant or undetected for hours, days, weeks, or even longer, until one day they adversely affect the unsuspecting aircrew. Consequently, investigators with even the best intentions may overlook them.

Within this concept of latent failures, Reason described three more levels of human failure that contribute to the breakdown of a productive system. The first level involves conditions that directly affect operator performance. Referred to as *preconditions for unsafe acts*, this level involves conditions such as mental fatigue or improper communication and coordination practices, often referred to as crew resource management (CRM). Predictably, if fatigued aircrew fail to communicate and coordinate their activities with others in the cockpit or individuals external to the aircraft (e.g., air traffic control, maintenance, etc.), poor decisions are made and errors often result.

But exactly why did communication and coordination break down in the first place? This is perhaps where Reason's work departs from traditional approaches to human error. In many instances, the breakdown in good CRM practices can be traced back to instances of *unsafe supervision*, the third level of human failure. If, for example, two inexperienced (and perhaps even, below average pilots) are paired with each other and sent on a flight into known adverse weather at night, is anyone really surprised by a tragic outcome? To make matters worse, if this questionable manning practice is coupled with the lack of quality CRM training, the potential for miscommunication and ultimately, aircrew errors, is magnified. In a sense then, the crew was "set up" for failure as crew coordination and ultimately performance would be compromised. This is not to lessen the role played by the aircrew, only that intervention and mitigation strategies might lie higher within the system.

Reason's model did not stop at the supervisory level either; the *organization* itself can impact performance at all levels. For instance, in times of fiscal austerity, cash is at a premium, and as a result, training and sometimes even flight time are dramatically reduced. Consequently, supervisors are often left with no alternative but to task "non-proficient" aviators with complex tasks. Not surprisingly then, in the absence of good CRM training, communication and coordination failures will begin to appear as will a myriad of other preconditions, all of which will affect performance and elicit aircrew errors. Therefore, it makes sense that, if the accident rate is going to be reduced beyond current levels, investigators and analysts alike must examine the accident sequence in its entirety and expand it beyond the cockpit. Ultimately, causal factors at all levels within the organization must be addressed if any accident investigation and prevention system is going to succeed.

Strengths and Limitations of Reason's Model

It is easy to see how Reason's "Swiss cheese" model of human error integrates the human error perspectives described in Chapter 2 into a single unified framework. For example, the model is based on the premise that aviation operations can be viewed as a complex productive system (ergonomic perspective), that often breaks down because of ill-fated decisions made by upper level management and supervisors (organizational perspective). However, the impact that these fallible decisions have on safe operations may lie dormant for long periods of time until they produce unsafe operating conditions, such as poorly maintained equipment (ergonomic perspective), as well as unsafe aircrew conditions, such as fatigue (aeromedical perspective) or miscommunications among operators (psychosocial perspective). All of these factors, in turn affect an operators' ability to process information and perform efficiently (cognitive perspective). The result is often "pilot error," followed by an incident or accident.

A limitation of Reason's model, however, is that it fails to identify the exact nature of the "holes" in the cheese. After all, as a safety officer or accident investigator, wouldn't you like to know what the holes in the "cheese" are? Wouldn't you like to know the types of organizational and supervisory failures that "trickle down" to produce failed defenses at the preconditions or unsafe acts level?

It should also be noted that the original description of his model was geared toward academicians rather than practitioners. Indeed, some have suggested that the unsafe acts level as described by Reason and others was too theoretical. As a result, analysts, investigators, and other safety professionals have had a difficult time applying Reason's model to the "real-world" of aviation.

This predicament is evidenced by ICAO's (1993) human-factors accident investigation manual. This manual describes Reason's model and touts it as a great advancement in our understanding of the human causes of aviation accidents. However, the manual then reverts to the SHEL model as a framework for investigating accidents. This is because Reason's model is primarily descriptive, not analytical. For the model to be systematically and effectively utilized as an analysis tool, the "holes in the cheese" need to be clearly defined. One needs to know what these system failures or "holes" are, so that they can be identified during accident investigations or better yet, detected and corrected before an accident occurs.

Defining the Holes in the Cheese: The Human Factors Analysis and Classification System (HFACS)

The Human Factors Analysis and Classification System (HFACS) was specifically developed to define the latent and active failures implicated in Reason's "Swiss cheese" model so it could be used as an accident investigation and analysis tool (Shappell and Wiegmann, 1997a; 1998; 1999; 2000a; 2001). The framework was developed and refined by analyzing hundreds of accident reports containing thousands of human causal factors. Although designed originally for use within the context of military aviation, HFACS has been shown to be effective within the civil aviation arena as well (Wiegmann and Shappell, 2001b). Specifically, HFACS describes four levels of failure, each of which corresponds to one of the four layers contained within Reason's model. These include: 1) Unsafe Acts, 2) Preconditions for Unsafe Acts, 3) Unsafe Supervision, and 4) Organizational Influences. The balance of this chapter describes the causal categories associated with each of these levels.

Unsafe Acts of Operators

The unsafe acts of operators can be loosely classified into two categories: errors and violations (Reason, 1990). In general, errors represent the mental or physical activities of individuals that fail to achieve their intended outcome. Not surprising, given the fact that humans by their very nature make errors, these unsafe acts dominate most accident databases. Violations, on the other hand, refer to the willful disregard for the rules and regulations that govern the safety of flight. The bane of many organizations, the prediction and prevention of these inexcusable and purely "preventable" unsafe acts, continue to elude managers and researchers alike.

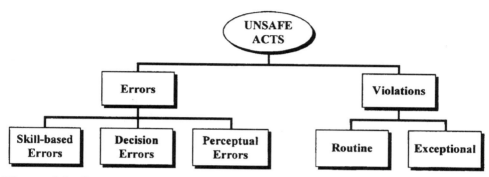

Figure 3.3 Categories of unsafe acts committed by aircrews

Still, distinguishing between errors and violations does not provide the level of granularity required of most accident investigations. Therefore, the categories of errors and violations were expanded here (Figure 3.3), as elsewhere (Reason, 1990; Rasmussen, 1982), to include three basic error types (skill-based, decision, and perceptual errors) and two forms of violations (routine and exceptional).

Errors

Skill-based errors. Skill-based behavior within the context of aviation is best described as "stick-and-rudder" and other basic flight skills that occur without significant conscious thought. As a result, these skill-based actions are particularly vulnerable to failures of attention and/or memory. In fact, attention failures have been linked to many skill-based errors such as the breakdown in visual scan patterns, task fixation, the inadvertent activation of controls, and the misordering of steps in a procedure, among others (Table 3.1). A classic example is an aircrew that becomes so fixated on troubleshooting a burnt out warning light that they do not notice their fatal descent into the terrain. Perhaps a bit closer to home, consider the hapless soul who locks himself out of the car or misses his exit because he was either distracted, in a hurry, or daydreaming. These are all examples of attention failures that commonly occur during highly automated behavior. While

Table 3.1   Selected examples of unsafe acts of operators

ERRORS	VIOLATIONS
Skill-based Errors	*Routine*
❑ Breakdown in visual scan	❑ Inadequate briefing for flight
❑ Inadvertent use of flight controls	❑ Failed to use ATC radar advisories
❑ Poor technique/airmanship	❑ Flew an unauthorized approach
❑ Over-controlled the aircraft	❑ Violated training rules
❑ Omitted checklist item	❑ Filed VFR in marginal weather conditions
❑ Omitted step in procedure	
❑ Over-reliance on automation	❑ Failed to comply with departmental manuals
❑ Failed to prioritize attention	
❑ Task overload	❑ Violation of orders, regulations, SOPs
❑ Negative habit	
❑ Failure to see and avoid	❑ Failed to inspect aircraft after in-flight caution light
❑ Distraction	
Decision Errors	*Exceptional*
❑ Inappropriate maneuver/procedure	❑ Performed unauthorized acrobatic maneuver
❑ Inadequate knowledge of systems, procedures	❑ Improper takeoff technique
	❑ Failed to obtain valid weather brief
❑ Exceeded ability	❑ Exceeded limits of aircraft
❑ Wrong response to emergency	❑ Failed to complete performance computations for flight
Perceptual Errors	
❑ *Due to* visual illusion	❑ Accepted unnecessary hazard
❑ *Due to* spatial disorientation/vertigo	❑ Not current/qualified for flight
	❑ Unauthorized low-altitude canyon running
❑ *Due to* misjudged distance, altitude, airspeed, clearance	

Note: This is not a complete listing.

these attention/memory failures may be frustrating at home or driving around town, in the air, they can become catastrophic.

In contrast to attention failures, memory failures often appear as omitted items in a checklist, place losing, or forgotten intentions. Indeed, these are common everyday occurrences for most of us. For example, who among us hasn't sent an email to someone with the intention of attaching a file, only to find out later that you forgot to attach the file? Likewise, many coffee drinkers have, at least one time in their life, brewed only water because they forgot to put coffee in the coffeemaker. If such errors can occur in seemingly benign situations such as these, it should come as no surprise that when under the stress of an inflight emergency, critical steps in emergency procedures can be missed. However, even when not particularly stressed, pilots have been known to forget to set the flaps on approach or lower the landing gear – at a minimum, an embarrassing gaffe.

The third, and final, type of skill-based errors identified in many accident investigations involves technique errors. Regardless of one's training, experience, and educational background, the manner in which one carries out a specific sequence of actions may vary greatly. That is, two pilots with identical training, flight grades, and experience may differ significantly in the manner in which they maneuver their aircraft. While one pilot may fly smoothly with the grace of a soaring eagle, others may fly with the darting, rough transitions of a sparrow. Although both may be safe and equally adept at flying, the techniques they employ could set them up for specific failure modes. In fact, such techniques are as much a factor of innate ability and aptitude as they are an overt expression of one's own personality, making efforts at the prevention and mitigation of technique errors difficult, at best.

Decision errors. The second error form, decision errors, represents intentional behavior that proceeds as planned, yet the plan itself proves inadequate or inappropriate for the situation (Table 3.1). Often referred to as "honest mistakes," these unsafe acts represent the actions or inactions of individuals whose "hearts are in the right place," but they either did not have the appropriate knowledge or just simply chose poorly.

Perhaps the most heavily investigated of all error forms, decision errors can be grouped into three general categories: procedural errors, poor choices, and problem-solving errors. Procedural decision errors (Orasanu, 1993), or rule-based mistakes as described by Rasmussen (1982), occur during highly structured tasks of the sorts, if X, then do Y. Aviation, particularly within the military and commercial environments, by its very nature is highly structured, and consequently, much of pilot decision-making is procedural. There are very explicit procedures to be performed at virtually all phases of flight. Still, errors can, and often do, occur when a situation is either not recognized or misdiagnosed, and the wrong procedure is applied. This is particularly true when pilots are placed in time-critical emergencies like an engine malfunction on takeoff.

However, even in aviation, not all situations have corresponding procedures to deal with them. Therefore, many situations require a choice to be made among multiple response options. Consider the pilot flying home after a long week away from the family who unexpectedly confronts a line of thunderstorms directly in his path. He can choose to fly around the weather, divert to another field until the weather passes, or penetrate the weather hoping to quickly transition through it. Confronted with situations such as this, choice decision errors (Orasanu, 1993), or knowledge-based mistakes as they are otherwise known (Rasmussen, 1982), may occur. This is particularly true when there is insufficient experience, time, or other outside pressures that may preclude safe decisions. Put simply, sometimes we chose well, and sometimes we do not.

Finally, there are occasions when a problem is not well understood, and formal procedures and response options are not available. It is during these ill-defined situations that the invention of a novel solution is required. In a sense, individuals find themselves where they have not been before, and in many ways, must literally fly by the seat of their pants. Individuals placed in this situation must resort to slow and effortful reasoning processes where time is a luxury rarely afforded. Not surprisingly, while this type of decision-making is more infrequent than other forms, the relative proportion of problem-solving errors committed is markedly higher.

Admittedly, there are a myriad of other ways to describe decision errors. In fact, numerous books have been written on the topic. However, the point here is that decision errors differ markedly from skill-based errors in that the former involve deliberate and conscious acts while the latter entail highly automatized behavior.

Perceptual errors. Predictably, when one's perception of the world differs from reality, errors can, and often do, occur. Typically, perceptual errors occur when sensory input is either degraded or "unusual," as is the case with visual illusions and spatial disorientation or when aircrews simply misjudge the aircraft's altitude, attitude, or airspeed (Table 3.1). Visual illusions, for example, occur when the brain tries to "fill in the gaps" with what it feels belongs in a visually impoverished environment, such as that seen at night or when flying in adverse weather. Likewise, spatial disorientation occurs when the vestibular system cannot resolve one's orientation in space and therefore makes a "best guess" when visual (horizon) cues are absent. In either event, the unsuspecting individual is often left to make a decision that is based on faulty information, and the potential for committing an error is elevated.

It is important to note, however, that it is not the illusion or disorientation that is classified as a perceptual error. Rather, it is the pilot's erroneous response to the illusion or disorientation. For example, many unsuspecting pilots have experienced "black-hole" approaches, only to fly a perfectly good aircraft into the terrain or water. This continues to occur, even though it is well known that flying at night over dark, featureless terrain (e.g., a lake or field devoid of trees), will produce the illusion that the aircraft is actually higher than it is. As a result, pilots are taught to rely on their primary instruments, rather than the outside world, particularly during the approach phase of flight. Even so, some pilots fail to monitor their instruments when flying at night. Tragically, these aircrew and others who have been fooled by illusions and other disorientating flight regimes may end up involved in a fatal aircraft accident.

Violations

By definition, errors occur within the rules and regulations espoused by an organization. In contrast, violations represent a willful disregard for the rules and regulations that govern safe flight and, fortunately, occur much less frequently since they often involve fatalities (Shappell et al., 1999).

Routine Violations. While there are many ways to distinguish between types of violations, two distinct forms have been identified, based on their etiology, that will help the safety professional when identifying accident causal factors. The first, routine violations, tend to be habitual by nature and often tolerated by governing authority (Reason, 1990). Consider, for example, the individual who drives consistently 5–10 mph faster than allowed by law or someone who routinely flies in marginal weather when authorized for visual flight rules (VFR) only. While both are certainly against existing regulations, many others have done the same thing. Furthermore, those who regularly drive 64 mph in a 55-mph zone, almost always drive 64 mph in a 55-mph zone. That is, they "routinely" violate the speed limit. The same can typically be said of the pilot who routinely flies into marginal weather.

What makes matters worse, these violations (commonly referred to as "bending" the rules) are often tolerated and, in effect, sanctioned by supervisory authority (i.e., you're not likely to get a traffic citation until you exceed the posted speed limit by more than 10 mph). If, however, the local authorities started handing out traffic citations for exceeding the speed limit on the highway by 9 mph or less (as is often done on military installations), then it is less likely that individuals would violate the rules. Therefore, by definition, if a routine violation is identified, one must look further up the supervisory chain to identify those individuals in authority who are not enforcing the rules.

Exceptional Violations. On the other hand, unlike routine violations, exceptional violations appear as isolated departures from authority, not necessarily indicative of an individual's typical behavior pattern, nor condoned by management (Reason, 1990). For example, an isolated instance of driving 105 mph in a 55-mph zone is considered an exceptional violation since it is highly unlikely that the individual does this all the time. Likewise, flying under a bridge or engaging in other prohibited maneuvers, like low-level canyon running, would constitute an exceptional violation. However, it is important to note that, while most exceptional violations are heinous, they are not considered "exceptional" because of their extreme nature. Rather, they are considered exceptional because they are neither typical of the individual, nor condoned by authority.

56 *A Human Error Approach to Aviation Accident Analysis*

Still, what makes exceptional violations particularly difficult for any organization to deal with is that they are not indicative of an individual's behavioral repertoire and, as such, are particularly difficult to predict. In fact, when individuals are confronted with evidence of their dreadful behavior and asked to explain it, they are often left with little explanation. Indeed, those individuals who survived such excursions from the norm clearly knew that, if caught, dire consequences would follow. Nevertheless, defying all logic, many otherwise model citizens have been down this potentially tragic road.

Preconditions for Unsafe Acts

Arguably, the unsafe acts of aircrew can be directly linked to nearly 80% of all aviation accidents. However, simply focusing on unsafe acts is like focusing on a fever without understanding the underlying illness that is causing it. Thus, investigators must dig deeper into why the unsafe acts occurred in the first place. The process involves analyzing preconditions of unsafe acts, which includes the condition of the operators, environmental and personnel factors (Figure 3.4).

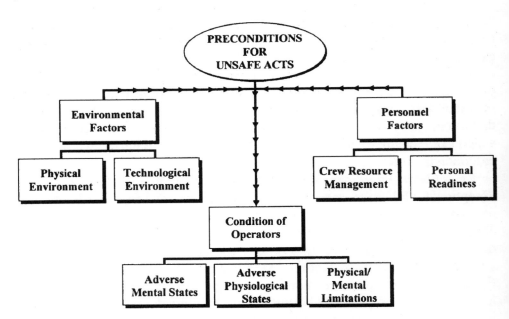

Figure 3.4 Categories of preconditions of unsafe acts

Condition of Operators

The condition of an individual can, and often does, influence performance on the job whether it is flying a plane, operating on a patient, or working on an assembly line. Unfortunately, this critical link in the chain of events leading up to an accident often goes unnoticed by investigators who have little formal training in human factors, psychology, or aerospace medicine. Still, it does not require a degree in any of those fields to thoroughly examine these potentially dangerous factors. Sometimes, it just takes pointing investigators in the right direction and letting their natural instincts take over. That is our purpose as we briefly describe three conditions of operators that directly impact performance: Adverse mental states, adverse physiological states, and physical/mental limitations.

Adverse mental states. Being prepared mentally is critical in nearly every endeavor, but perhaps even more so in aviation. As such, the category of adverse mental states was created to account for those mental conditions that affect performance (Table 3.2). Principal among these are the loss of situational awareness, task fixation, distraction, and *mental* fatigue due to sleep loss or other stressors. Also included in this category are personality traits and pernicious attitudes such as overconfidence, complacency, and misplaced motivation.

Predictably, if an individual is mentally tired for whatever reason, the likelihood that an error will occur increases. In a similar fashion, overconfidence and other hazardous attitudes such as arrogance and impulsivity will influence the likelihood that a violation will be committed. Clearly then, any framework of human error must account for these preexisting adverse mental states in the causal chain of events.

Adverse physiological states. The second category, adverse physiological states, refers to those medical or physiological conditions that preclude safe operations (Table 3.2). Particularly important to aviation are such conditions as visual illusions and spatial disorientation as described earlier, as well as *physical* fatigue and the myriad of pharmacological and medical abnormalities known to affect performance.

The effects of visual illusions and spatial disorientation are well known to most aviators. However, the effects on cockpit performance of simply being ill are less well known and often overlooked. Nearly all of us have gone to work sick, dosed with over-the-counter medications, and have generally performed well. Consider however, the pilot suffering from the common head cold. Unfortunately, most aviators view a head cold as only a minor inconvenience that can be easily remedied using over-the-counter antihistamines, acetaminophen, and other non-prescription pharmaceuticals. In fact, when confronted with a stuffy nose, aviators typically are only concerned with the effects of a painful sinus block as cabin altitude changes.

Table 3.2 Selected examples of preconditions of unsafe acts

CONDITION OF OPERATOR

Adverse Mental States
- Loss of situational awareness
- Complacency
- Stress
- Overconfidence
- Poor flight vigilance
- Task saturation
- Alertness (drowsiness)
- Get-home-itis
- Mental fatigue
- Circadian dysrhythmia
- Channelized attention
- Distraction

Adverse Physiological States
- Medical illness
- Hypoxia
- Physical fatigue
- Intoxication
- Motion sickness
- Effects of OTC medications

Physical/Mental Limitations
- Visual limitations
- Insufficient reaction time
- Information overload
- Inadequate experience for complexity of situation
- Incompatible physical capabilities
- Lack of aptitude to fly
- Lack of sensory input

PERSONNEL FACTORS

Crew Resource Management
- Failed to conduct adequate brief
- Lack of teamwork
- Lack of assertiveness
- Poor communication/ coordination within & between aircraft, ATC, etc.
- Misinterpretation of traffic calls
- Failure of leadership

Personal Readiness
- Failure to adhere to crew rest requirements
- Inadequate training
- Self-medicating
- Overexertion while off duty
- Poor dietary practices
- Pattern of poor risk judgment

ENVIRONMENTAL FACTORS

Physical Environment
- Weather
- Altitude
- Terrain
- Lighting
- Vibration
- Toxins in the cockpit

Technological Environment
- Equipment/controls design
- Checklist layout
- Display/interface characteristics
- Automation

Note: This is not a complete listing.

But, it is not the overt symptoms that concern the local flight surgeon. Rather, it is the potential inner ear infection and increased likelihood of spatial disorientation while flying in IMC that alarms them – not to mention the fatigue and sleep loss that often accompany an illness. Therefore, it is incumbent upon any safety professional to account for these sometimes subtle, yet potentially harmful medical conditions when investigating an accident or incident.

Physical/Mental Limitations. The third and final category involves an individual's physical/mental limitations (Table 3.2). Specifically, this category refers to those instances when operational requirements exceed the capabilities of the individual at the controls. For example, the human visual system is severely limited at night; yet, automobile drivers do not necessarily slow down or take additional precautions while driving in the dark. In aviation, while slowing down is not really an option, paying additional attention to basic flight instruments and increasing one's vigilance will often add to the safety margin. Unfortunately, when precautions are not taken, the results can be catastrophic, as pilots will often fail to see other aircraft, obstacles, or power lines due to the size or contrast of the object in the visual field.

There are also occasions when the time required to complete a task or maneuver exceeds one's ability. While individuals vary widely in their capacity to process and respond to information, pilots are typically noted for their ability to respond quickly. But faster does not always mean better. It is well documented, that if individuals are required to respond quickly (i.e., less time is available to consider all the possibilities or choices thoroughly), the probability of making an error goes up markedly. It should be no surprise then, that when faced with the need for rapid processing and reaction time, as is the case in many aviation emergencies, all forms of error would be exacerbated.

Perhaps more important than these basic sensory and information processing limitations, there are at least two additional issues that need to be addressed – albeit they are often overlooked or avoided for political reasons by many safety professionals. These involve individuals who simply are not compatible with aviation, because they are either unsuited physically or do not possess the aptitude to fly. For instance, some people simply do not have the physical strength to operate in the potentially high-G environment of military or aerobatic aviation, or for anthropometric reasons, simply have difficulty reaching the controls or seeing out the windscreen. In other words, cockpits have traditionally not been designed with all shapes, sizes, and physical abilities in mind. Indeed, most cockpits have been designed around the average male, making flying particularly difficult for those less than 5 feet tall or over 6.5 feet tall.

A much more sensitive topic to address as an accident investigator is the fact that not everyone has the mental ability or aptitude to fly. Just as not all of us can be concert pianists or NFL linebackers, not everyone has the innate ability to pilot an aircraft – a vocation that requires the unique ability to make decisions quickly, on limited information, and *correct the first time* in life-threatening situations. This does not necessarily have anything to do with raw intelligence. After all, we have argued that the eminent Albert Einstein would likely not have been a good pilot because, like many scientists, he always looked for the perfect answer, a luxury typically not afforded during an in-flight emergency. The difficult task for the safety professional is identifying whether aptitude might have contributed to the accident.

Personnel Factors

It is not difficult to envision how the condition of an operator can lead to the commission of unsafe acts. Nevertheless, there are a number of things that aircrew often do to themselves to create these preconditions for unsafe acts. We like to refer to them as personnel factors, and have divided them into two general categories: crew resource management and personal readiness.

Crew Resource Management. Good communication skills and team coordination have been the mantra of I/O and personnel psychology for years. Not surprising then, crew resource management has been a cornerstone of aviation as well (Helmreich and Foushee, 1993). As a result, this category was created to account for occurrences of poor coordination among personnel (Table 3.2). Within the context of aviation, this includes coordination within and between aircraft, as well as with air traffic control, maintenance, and other support personnel. But aircrew coordination does not stop with the aircrew in flight. It also includes coordination before takeoff and after landing with the brief and debrief of the aircrew.

It is not difficult to envision a scenario where the lack of crew coordination has led to confusion and poor decision-making in the cockpit. In fact, aviation accident databases are littered with instances of poor coordination among aircrew. One of the more tragic examples was the crash of a civilian airliner at night in the Florida Everglades as the crew was busily trying to troubleshoot what amounted to little more than a burnt out indicator light. Unfortunately, no one in the cockpit was monitoring the aircraft's altitude as the autopilot was inadvertently disconnected. Ideally, the crew would have coordinated the troubleshooting task ensuring that at least one crewmember was monitoring basic flight instruments and "flying" the aircraft. Tragically, this was not the case, as they entered a slow, unrecognized descent into the swamp resulting in numerous fatalities.

Personal Readiness. In aviation, or for that matter in any occupational setting, individuals are expected to show up for work ready to perform at optimal levels. A breakdown in personal readiness can occur when individuals fail to prepare physically or mentally for duty (Table 3.2). For instance, violations of crew rest requirements, bottle-to-brief rules, and self-medicating will all affect performance on the job and are particularly detrimental in the aircraft. For instance, it is not hard to imagine that when individuals do not adhere to crew rest requirements, that they run the risk of suffering from mental fatigue and other adverse mental states, which ultimately lead to errors and accidents. Note however, that violations that affect personal readiness are not considered an "unsafe act, violation" since they typically do not happen in the cockpit, nor are they necessarily active failures with direct and immediate consequences.

Still, not all personal readiness failures occur because rules or regulations have been broken. For example, jogging 10 miles before piloting an aircraft may not be against any existing regulations, yet it may impair the physical and mental capabilities of the individual enough to degrade performance and elicit unsafe acts. Likewise, the traditional "candy bar and coke" lunch of the military pilot may sound good, but is often not enough to sustain performance in the demanding environment of aviation. While there may be no rules governing such behavior, pilots must use good judgment when deciding whether they are "fit" to fly an aircraft.

Environmental Factors

In addition to personnel factors, environmental factors can also contribute to the substandard conditions of operators and hence to unsafe acts. Very broadly, these *environmental factors* can be captured within two general categories: the physical environment and the technological environment.

Physical environment. The impact that the physical environment can have on aircrew has long been known and much has been documented in the literature on this topic (e.g., Nicogossian et al., 1994; Reinhart, 1996). The term physical environment refers to both the operational environment (e.g., weather, altitude, terrain), and the ambient environment, such as heat, vibration, lighting, toxins, etc. in the cockpit (Table 3.2). For example, as mentioned earlier, flying into adverse weather reduces visual cues, which can lead to spatial disorientation and perceptual errors. Other aspects of the physical environment, such as heat, can cause dehydration that reduces a pilot's concentration level, producing a subsequent slowing of decision-making processes or even the inability to control the aircraft. In military aircraft, and even occasionally during aerobatic flight in civil aircraft, acceleration forces can cause a restriction in blood flow to the brain, producing blurred vision or even unconsciousness. Furthermore, a loss of pressurization at high altitudes, or maneuvering at high altitudes without supplemental oxygen in unpressurized aircraft, can obviously result in hypoxia, which leads to delirium, confusion, and a host of unsafe acts.

Technological environment. The technological environment that pilots often find themselves in can also have a tremendous impact on their performance. While the affect of some of these factors has been known for a long time, others have only recently received the attention they deserve. Within the context of HFACS, the term *technological environment* encompasses a variety of issues including the design of equipment and controls, display/interface characteristics, checklist layouts, task factors and automation (Table 3.2). For example, one of the classic design problems first discovered in aviation was the similarity between the controls used to raise and lower the flaps and those used to raise and lower the landing gear. Such similarities often caused confusion among pilots, resulting in the frequent raising of the landing gear while still on the ground. Needless to say, this made a seemingly routine task like taxiing for take-off much more exciting! A more recent problem with cockpit interfaces is the method used in some aircraft to communicate the location of a particular engine failure. Many of us have likely read accident reports or heard about pilots who experienced an engine failure in-flight and then inadvertently shut down the wrong engine, leaving them without a good propulsion system – an unenviable situation for any pilot to be in. After all, there is no worse feeling as a pilot than to be in a glider that only moments early was a powered airplane.

The redesign of aircraft systems and the advent of more complex glass-cockpits have helped reduce a number of these problems associated with human error. However, they have also produced some new problems of their own. For example, human–automation interactions are extremely complex and frequently reveal nuances in human behavior that no one anticipated. Highly reliable automation, for instance, has been shown to induce adverse mental states such as over-trust and complacency, resulting in pilots following the instructions of the automation even when "common sense" suggests otherwise. In contrast, imperfectly reliable automation can often result in under-trust and disuse of automation even though aided performance is safer than unaided performance (Wickens and Hollands, 2000). Pilots turning off their traffic collision avoidance system (TCAS) because it often produces false alarms would be one example.

In other cases, the interfaces associated with the automation can produce problems, such as the multiple modes associated with modern flight management systems (FMS). Pilots often suffer from "mode confusion" while interacting with these systems (Sarter and Woods, 1992). As a result, they may make dire decision errors and subsequently fly a perfectly good aircraft into the ground.

Unsafe Supervision

Recall that Reason's (1990) "Swiss cheese" model of accident causation includes supervisors who influence the condition of pilots and the type of environment they operate in. As such, we have identified four categories of unsafe supervision: Inadequate supervision, planned inappropriate operations, failure to correct a known problem, and supervisory violations (Figure 3.5). Each is described briefly below.

Figure 3.5 Categories of unsafe supervision

Inadequate Supervision. The role of any supervisor is to provide their personnel the opportunity to succeed. To do this, they must provide guidance, training, leadership, oversight, incentives, or whatever it takes, to ensure that the job is done safely and efficiently. Unfortunately, this is not always easy, nor is it always done. For example, it is not difficult to conceive of a situation where adequate crew resource management training was either not provided, or the opportunity to attend such training was not afforded to a particular aircrew member. As such, aircrew coordination skills would likely be compromised, and if the aircraft were put into an adverse situation (an emergency, for instance), the risk of an error being committed would be exacerbated and the potential for an accident would increase significantly.

In a similar vein, sound professional guidance and oversight are essential ingredients in any successful organization. While empowering individuals to make decisions and function independently is certainly important, this does not divorce the supervisor from accountability. The lack of guidance and oversight has proven to be a breeding ground for many of the violations that have crept into the cockpit. As such, any thorough investigation of accident causal factors must consider the role supervision plays (i.e., whether the supervision was inappropriate or did not occur at all) in the genesis of human error (Table 3.3).

Planned Inappropriate Operations. Occasionally, the operational tempo and/or the scheduling of aircrew is such that individuals are put at unacceptable risk, crew rest is jeopardized, and ultimately performance is adversely affected. Such operations, though arguably unavoidable during emergencies, are otherwise regarded as unacceptable. Therefore, the second category of unsafe supervision, planned inappropriate operations, was created to account for these failures (Table 3.3).

Table 3.3   Selected examples of unsafe supervision

Inadequate Supervision
- Failed to provide proper training
- Failed to provide professional guidance/oversight
- Failed to provide current publications/adequate technical data and/or procedures
- Failed to provide adequate rest period
- Lack of accountability
- Perceived lack of authority
- Failed to track qualifications
- Failed to track performance
- Failed to provide operational doctrine
- Over-tasked/untrained supervisor
- Loss of supervisory situational awareness

Planned Inappropriate Operations
- Poor crew pairing
- Failed to provide adequate brief time/supervision
- Risk outweighs benefit
- Failed to provide adequate opportunity for crew rest
- Excessive tasking/workload

Failed to Correct a Known Problem
- Failed to correct inappropriate behavior/identify risky behavior
- Failed to correct a safety hazard
- Failed to initiate corrective action
- Failed to report unsafe tendencies

Supervisory Violations
- Authorized unqualified crew for flight
- Failed to enforce rules and regulations
- Violated procedures
- Authorized unnecessary hazard
- Willful disregard for authority by supervisors
- Inadequate documentation
- Fraudulent documentation

Note: This is not a complete listing.

Consider, for example, the issue of improper crew pairing. It is well known that when very senior, dictatorial captains are paired with very junior, weak co-pilots, communication and coordination problems are likely to occur. Commonly referred to as the trans-cockpit authority gradient, such conditions likely contributed to the tragic crash of a commercial airliner into the Potomac River outside of Washington, DC, in January of 1982 (NTSB,

1982). In that accident, the captain of the aircraft repeatedly rebuffed the first officer when the latter indicated that the engine instruments did not appear normal. Undaunted, the captain continued a fatal takeoff in icing conditions with less than adequate takeoff thrust. Tragically, the aircraft stalled and plummeted into the icy river, killing the crew, and many of the passengers.

Obviously, the captain and crew were held accountable – after all, they tragically died in the accident. But what was the role of the supervisory chain? Perhaps crew pairing was equally responsible. Although not specifically addressed in the report, such issues are clearly worth exploring in many accidents. In fact, in this particular instance, several other training and manning issues were also identified.

Failure to Correct a Known Problem. The third category, failure to correct a known problem, refers to those instances when deficiencies among individuals, equipment, training or other related safety areas are "known" to the supervisor, yet are allowed to continue unabated (Table 3.3). For example, it is not uncommon for accident investigators to interview a pilot's friends, colleagues, and supervisors after a fatal crash only to find out that they "knew it would happen to him some day." If the supervisor knew that a pilot was incapable of flying safely, and allowed the flight anyway, he clearly did the pilot no favors. Some might even say that the failure to correct the behavior, either through remedial training or, if necessary, removal from flight status, essentially signed the pilot's death warrant – not to mention that of others who may have been on board.

Likewise, the failure to consistently correct or discipline inappropriate behavior certainly fosters an unsafe atmosphere and promotes the violation of rules. Aviation history is rich with reports of aviators who tell hair-raising stories of their exploits and barnstorming low-level flights (the infamous "been there, done that"). While entertaining to some, they often serve to promulgate a perception of tolerance and "one-up-manship" until one day someone ties the low altitude flight record of ground-level! Indeed, the failure to report these unsafe tendencies and initiate corrective actions is yet another example of the failure to correct known problems.

Supervisory Violations. Supervisory violations, on the other hand, are reserved for those instances when existing rules and regulations are willfully disregarded by supervisors (Table 3.3). Although arguably rare, supervisors have been known to occasionally violate the rules and doctrine when managing their assets. For instance, there have been occasions when individuals were permitted to operate an aircraft without current qualifications or license. Likewise, it can be argued that failing to enforce existing rules and regulations or flaunting authority are also violations at the supervisory level. While rare and possibly difficult to cull out, such practices are a flagrant violation of the rules and invariably set the stage for the tragic sequence of events that predictably follow.

Organizational Influences

As noted previously, fallible decisions of upper-level management can directly affect supervisory practices, as well as the conditions and actions of operators. Unfortunately, these organizational errors often go unnoticed by safety professionals, due in large part to the lack of a clear framework from which to investigate them. Generally speaking, the most elusive latent failures revolve around issues related to resource management, organizational climate, and operational processes, as detailed below and illustrated in Figure 3.6.

Figure 3.6 Organizational factors influencing accidents

Resource Management. This category encompasses the realm of corporate-level decision-making regarding the allocation and maintenance of organizational assets such as human resources (personnel), monetary assets, equipment, and facilities (Table 3.4). Generally speaking, corporate decisions about how such resources should be managed are typically based upon two, sometimes conflicting, objectives – the goal of safety and the goal of on-time, cost-effective operations. In times of relative prosperity, both objectives can be easily balanced and satisfied in full. However, as we mentioned earlier, there may also be times of fiscal austerity that demand some give-and-take between the two. Unfortunately, history tells us that safety and training are often the losers in such battles, and as such, the first to be cut in organizations having financial difficulties.

Excessive cost-cutting could also result in reduced funding for new equipment, the purchase of low-cost, less effective alternatives, or worse yet, the lack of quality replacement parts for existing aircraft and support equipment. Consider this scenario recently played out in the military. While waiting on a back-ordered part, one of the squadron's aircraft is parked in the hangar in a down status. In the meantime, other aircraft in the squadron suffer failures to parts that are also not readily available from supply. Naturally, the creative maintenance officer orders that parts be scavenged from the aircraft in the hangar (facetiously referred to as the "hangar queen") and put on the other jets on the flightline to keep them fully operational. The problem is that the "hangar queen" has to be flown every few weeks if the squadron is going to be able to maintain its high level of readiness (albeit only on paper). So, the parts are taken off the aircraft on the line, put on the "hangar queen" so it can be flown around the pattern a couple of times. Then, the parts are taken off the "hangar queen," put back on the other aircraft and the process continues until replacement parts arrive, or something worse happens. Alas, most aircraft parts are not designed to be put on and taken off, repeatedly. Soon, the inevitable occurs and a critical part fails in flight causing an accident. As accident investigators, do we consider this unconventional approach to maintenance as causal to the accident? Certainly, but the lack of readily available replacement parts because of poor logistics and resource management within the organization is equally culpable.

Organizational Climate. Organizational climate refers to a broad class of variables that influence worker performance (Table 3.4). Formally, it can be defined as the "situationally based consistencies in the organization's treatment of individuals" (Jones, 1988). While this may sound like psycho-babble to some, what it really means is that organizational climate can be viewed as the working atmosphere within the organization. One telltale sign of an organization's climate is its structure, as reflected in the chain-of-command, delegation of authority, communication channels, and formal accountability for actions. Just like in the cockpit, communication and coordination are also vital within an organization. If management and staff are not communicating, or if no one knows who is in charge, organizational safety clearly suffers and accidents can and will happen (Muchinsky, 1997).

An organization's culture and policies are also important variables related to climate. Culture really refers to the unofficial or unspoken rules, values, attitudes, beliefs, and customs of an organization. Put simply, culture is "the way things really get done around here." In fact, you will see in Chapter 5 how the culture within the U.S. Navy/Marine Corps actually contributed to a number of Naval aviation accidents.

Policies, on the other hand, are official guidelines that direct management's decisions about such things as hiring and firing, promotion, retention, sick leave, and a myriad of other issues important to the everyday business of the organization. When policies are ill-defined, adversarial, or conflicting, or when they are supplanted by unofficial rules and values, confusion abounds. Indeed, it is often the "unwritten policies" that are more interesting to accident investigators than the official ones. After all, it is safe to say that all commercial airlines have written policies on file that enable aircrew to request a relief pilot in the event they are too tired or ill to fly. While such policies exist on paper, there are some airlines whose "unwritten policies" make utilizing the relief pilot option difficult, and even career threatening in some instances. In fact, it can be argued that some corporate managers are quick to pay "lip service" to official safety policies while in the public eye, but then overlook such policies when operating behind the scenes.

Organizational Process. This category refers to corporate decisions and rules that govern the everyday activities within an organization, including the establishment and use of standard operating procedures and formal methods for maintaining checks and balances (oversight) between the workforce and management (Table 3.4). Consider, for example, a young and inexperienced aircraft mechanic right out of school tasked with changing an engine on a military fighter aircraft. As he dutifully lays out his manual and begins changing the engine, following the procedures step-by-step, along comes the salty old crew chief with 25 years of experience in the field. During the ensuing conversation, the chief is heard to say, "Son, if you follow that book, we'll never get this finished on time. Let me show you how it's done." Unfortunately, rather than follow the procedures as outlined in the manual, the chief relies more on his own experiences and memory than on the actual procedures in the manual. Perhaps the procedures themselves are faulty and there is no way that an engine can be changed in the time allowed when using the manual. Nevertheless, the non-standard procedure the chief is using also introduces unwanted variability into the maintenance operation. While the latter requires a different sort of remedial action, the former implies that the procedures themselves may be flawed and points toward a failure within the organizational process.

Other organizational factors such as operational tempo, time pressure, and work schedules are all variables that can adversely affect safety. As stated earlier, there may be instances when those within the upper echelon of an organization determine that it is necessary to increase the operational tempo to a point that overextends a supervisor's staffing capabilities. Therefore, a supervisor may have no recourse other than to utilize inadequate scheduling procedures that jeopardize crew rest or produce sub-optimal crew pairings, putting aircrew at an increased risk of a mishap. Clearly, organizations should have official procedures in place to address such contingencies, as well as oversight programs to monitor the risks. Regrettably, however, not all organizations have these procedures nor do they engage in an active process of monitoring aircrew errors and human factor problems via anonymous reporting systems and safety audits. As such, supervisors and managers are often unaware of the problems before an accident occurs.

Table 3.4 Selected examples of organizational influences

Resource Management

Human Resources
- ☐ Selection
- ☐ Staffing/manning
- ☐ Training
- ☐ Background checks

Monetary/Budget Resources
- ☐ Excessive cost cutting
- ☐ Lack of funding

Equipment/Facility Resources
- ☐ Poor aircraft/aircraft cockpit design
- ☐ Purchasing of unsuitable equipment
- ☐ Failure to correct known design flaws

Organizational Climate

Structure
- ☐ Chain-of-command
- ☐ Communication
- ☐ Accessibility/visibility of supervisor
- ☐ Delegation of authority
- ☐ Formal accountability for actions

Policies
- ☐ Promotion
- ☐ Hiring, firing, retention
- ☐ Drugs and alcohol
- ☐ Accident investigations

Culture
- ☐ Norms and rules
- ☐ Organizational customs
- ☐ Values, beliefs, attitudes

Organizational Process

Operations
- ☐ Operational tempo
- ☐ Incentives
- ☐ Quotas
- ☐ Time pressure
- ☐ Schedules

Procedures
- ☐ Performance standards
- ☐ Clearly defined objectives
- ☐ Procedures/instructions about procedures

Oversight
- ☐ Established safety programs/risk management programs
- ☐ Management's monitoring and checking of resources, climate, and processes to ensure a safe work environment

Note: This is not a complete listing.

Conclusion

Reason's Swiss cheese model provides a comprehensive theory of human error and accident causation. The Human Factors Analysis and Classification System (HFACS) was designed to define the "holes in the Swiss cheese" and to facilitate the application of this model to accident investigation and analysis. The purpose of this chapter, therefore, was to provide the reader with an overview of the categories contained within the framework. Figure 3.7 provides an illustration of all of these categories put together. We would like to emphasize that these categories were not just pulled out of thin air, or for that matter, out of a magician's hat. Nor were they developed only through brainstorming sessions with "expert" investigators. Rather, they were empirically derived and refined by analyzing hundreds of military and civil aviation accident reports that literally contained thousands of human causal factors. Still, HFACS must prove useful in the operational setting if it is to have any impact on aviation safety. Therefore, in the following chapters, we will demonstrate how HFACS can be used to investigate and analyze aviation accidents, as well as the new insights that can be gleaned from its application.

The Human Factors Analysis and Classification System 71

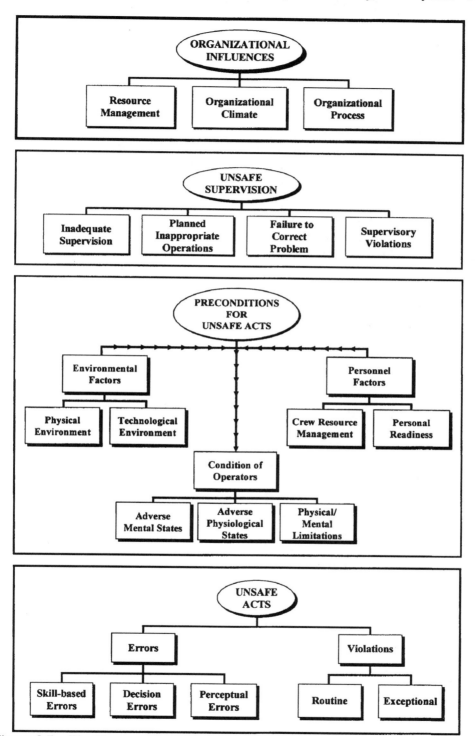

Figure 3.7   The Human Factors Analysis and Classification System (HFACS)

4 Aviation Case Studies using HFACS

To illustrate how HFACS can be used as an investigative as well as an analytical tool, we have chosen three U.S. commercial aviation accidents as case studies. For each, the final report made public by the National Transportation Safety Board (NTSB) was used as a resource. Often referred to as "blue covers" because of the color of ink used by the NTSB on their aviation accident report covers, these documents are quite detailed and represent the official findings, analyses, conclusions and recommendations of the NTSB. The interested reader can find a variety of these accident reports on the official NTSB web site (www.ntsb.gov) while many others (like the ones in this chapter) are available upon request. Interested readers can request the accidents described in this chapter and numerous others by writing to the NTSB at: National Transportation Safety Board, Public Inquiries Section, RE-51, 490 L'Enfant Plaza, S.W., Washington, D.C. 20594. Recently, Embry-Riddle Aeronautical University in Daytona Beach, Florida has scanned all the reports since 1967 and posted them in PDF format on their web site at (http://amelia.db.erau).

Before we begin, a word of caution is in order. It is possible that some readers may have more information regarding one or more of the accidents described below, or might even disagree with the findings of the NTSB, our analysis, or both. Nevertheless, our goal was not to reinvestigate the accident. To do so would be presumptuous and only infuse unwanted opinion, conjecture, and guesswork into the analysis process, since we were not privy to all the facts and findings of the case. Instead, we used *only* those causal factors determined by the NTSB, as well as the analyses contained within the report, when examining the accidents that follow.

It is also important to note that these case studies are presented for illustrative purposes only. While we have attempted to maintain the spirit and accuracy of the NTSB's analyses by citing specific findings and analyses (excerpts have been identified with italics and page number citations), in the interest of time we have only presented those details necessary to support the causes and contributing factors associated with each accident. Other information relevant to a thorough investigation (e.g., statements of fact), but not the cause of the accident, was excluded.

Sometimes Experience does Count

On a clear night in February of 1995, a crew of three positioned their DC-8 freighter with only three operable engines on the runway for what was intended to be a ferry flight from Kansas City, Missouri to Chicopee, Massachusetts for repairs. Just moments earlier the crew had aborted their initial takeoff attempt after losing directional control of the aircraft. Shortly after beginning their second attempt at the three-engine takeoff, the crew once again lost directional control and began to veer off the runway. This time however, rather than abort the takeoff as they had before, the captain elected to continue and rotated the aircraft early. As the aircraft briefly became airborne, it began an uncommanded roll to the left and crashed into the ground. Tragically, all three crewmembers were killed as the aircraft was destroyed (NTSB, 1995a).

The accident aircraft had arrived in Kansas City the day before as a regularly scheduled cargo flight from Denver, Colorado. The plan was to load it with new cargo and fly it to Toledo, Ohio later that day. However, during the engine start-up sequence, a failure in the gearbox prevented the No. 1 engine from starting. As luck would have it, the gearbox could not be repaired locally, so a decision was made to unload the cargo from the aircraft and fly it using its three good engines to Chicopee, Massachusetts the next day where repairs could be made.

Meanwhile, the accident aircrew was off-duty in Dover, Delaware having just completed a demanding flight schedule the previous two days. Their schedule called for them to spend the night in Dover, then ferry another aircraft to Kansas City the next afternoon. Because the crew would be in Kansas City, it was decided that they would be assigned the three-engine ferry flight to Chicopee – a somewhat surprising decision given that other crews, with more experience in three-engine takeoffs, would also be available.

Nevertheless, the decision was made, and later that night the chief pilot contacted the captain of the accident crew to discuss the three-engine ferry flight. Among the things discussed, were the weather forecast and a landing curfew of 2300 at Chicopee. Absent from the conversation however, was a review of three-engine takeoff procedures with the captain.

After an uneventful ferry flight from Dover to Kansas City the next day, the crew prepared for the three-engine ferry flight to Chicopee. Because the flight was expected to take a little over 2 hours, it was determined that they would have to depart Kansas City before 8 PM CST that night to arrive at

Chicopee before the airfield closed at 11 o'clock EST (note that there is a 1 hour time difference between Kansas City and Chicopee). Despite the curfew however, the engines were not even started until just past 8 PM CST, and then only after being interrupted by several procedural errors committed by the crew.

So already running a little bit late, the captain began to taxi the aircraft out to the runway and informed the crew that once airborne, they would need to fly as direct as possible to arrive at Chicopee before the 11 PM EST curfew. He also briefed the three-engine takeoff procedure; a brief that was characterized by a poor understanding of both three-engine throttle technique and minimum controllable airspeed during the takeoff ground run (Vmcg). After moving into position onto the runway and performing a static run-up of the three operable engines, the aircraft finally began its takeoff roll at roughly 8:20 PM CST (some 20 minutes behind schedule). However, during the takeoff roll, the power on the asymmetrical engine (in this case the number 4 engine, directly opposite the malfunctioning number 1 engine) was increased too rapidly, resulting in asymmetric thrust, causing the aircraft to veer to the left (Figure 4.1). It was at this point that the captain elected to abort the takeoff and taxi clear of the runway.

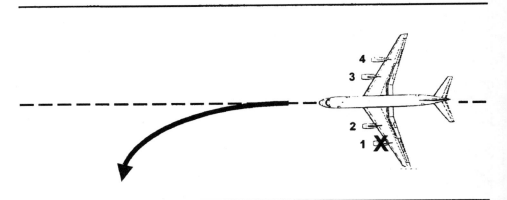

Note: The drawing is not to scale.

Figure 4.1 DC-8 with engine number 1 inoperable (marked with an "X") veers left due to asymmetrical thrust from number 4 engine

While taxiing back into position on the runway for another takeoff attempt, the crew discussed the directional control problems and its relationship to Vmcg. In particular, the crew focused on the rate at which power was to be applied to the asymmetrical (number 4) engine. Once again, the discussion was marked by confusion among the aircrew. This time however, the captain elected to depart from existing company procedures and let the flight engineer, rather than himself, control the power on the asymmetrical engine during the next takeoff attempt. This was done presumably so the captain could devote his attention to maintaining directional control of the aircraft.

Now, nearly 30 minutes behind schedule, the aircraft was once more positioned on the runway and cleared for takeoff. So, with the flight engineer at the throttles, the aircraft began its takeoff roll. However, this time the power was increased on the asymmetrical engine even faster than before. As a result, shortly after the first officer called "airspeed alive" at about 60 knots, the aircraft started an abrupt turn to the left followed quickly by a correction to the right. Then, almost immediately after the first officer called "90 knots,'" the aircraft once again started to veer to the left. However, unlike the initial takeoff attempt, the captain elected to continue, rather than abort, the takeoff. Seeing that the aircraft was going to depart the runway, the captain pulled back on the yoke and attempted to rotate the aircraft early (roughly 20 knots below the necessary speed to successfully takeoff). As the first officer called "we're off the runway," the aircraft briefly became airborne, yawed, and entered a slow 90-degree roll until it impacted the ground.

Human Factors Analysis using HFACS

There are many ways to conduct an HFACS analysis using NTSB reports. However, we have found that it is usually best to begin as investigators in the field do and work backward in time from the accident. In a sense then, we will roll the videotape backwards and conduct our analysis systematically. Using this approach, it was the captain's decision to continue the takeoff and rotate the aircraft some 20 knots below the computed rotation speed that ultimately sealed the crew's fate. That is to say, by continuing, rather than aborting the takeoff, the crew experienced a further loss of control, leading to a collision with the terrain. Hindsight being 20/20, what the captain should have done was abort the second takeoff as he had earlier, and made yet another attempt. Perhaps even better, the captain could have cancelled the flight altogether since it was unlikely that they would have arrived at Chicopee before the 2300 curfew anyway.

76 A Human Error Approach to Aviation Accident Analysis

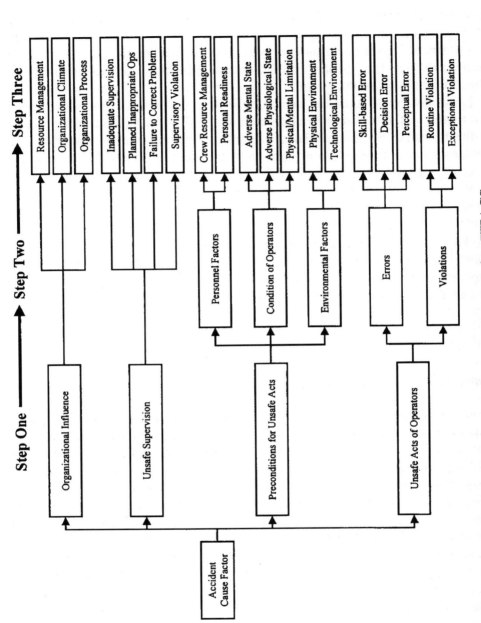

Figure 4.2 Steps required to classify causal factors using HFACS

So how did we classify the fact that the pilot elected to *"continue [rather than abort] the takeoff"* (NTSB, 1995a, p. 79) using HFACS? The classification process is really a two, or three-step process, depending on which level (i.e., unsafe acts, preconditions for unsafe acts, unsafe supervision, or organizational influences) you are working with (Figure 4.2). For instance, in this case, the decision to continue the takeoff was clearly an unsafe act. The next step involves determining whether the unsafe act was an error or violation. While certainly unwise, we have no evidence that the captain's decision violated any rules or regulations (in which case it would have been considered a violation). We therefore, classified it as an error.

Next, one has to determine which type of error (skill-based error, decision error, or perceptual error) was committed. Using definitions found in the previous chapter, it is unlikely that this particular error was the result of an illusion or spatial disorientation; hence, it was not considered a perceptual error. Nor was it likely the result of an automatized behavior/skill, in which case we would have classified it a skill-based error. Rather, the choice to continue the takeoff was a conscious decision on the part of the captain and therefore classified as a ***decision error***, using the HFACS framework.

Continuing with the analysis,[1] the next question one needs to ask is, "Why did the crew twice lose control of the aircraft on the takeoff roll?" The textbook answer is rather straightforward. By increasing the thrust on the asymmetric engine too quickly (i.e., more thrust from the engines attached to the right wing than the left), the captain on the first attempt and the flight engineer on the second, inadvertently caused the aircraft to drift left resulting in a loss of control. In both instances, an unsafe act was committed, in particular, an error: but, what type? The answer may reside with the crew's inexperience conducting three-engine takeoffs. It turns out that while all three crewmembers had completed simulator training on three-engine takeoffs the previous year, only the captain had actually performed one in an aircraft – and then only a couple of times as the first officer. When that inexperience was coupled with no real sense of how the airplane was responding to the increased thrust, an error was almost inevitable. Therefore, it was concluded that the manner in which the flight engineer applied thrust to the asymmetric engine led to the *"loss of directional control ... during the takeoff roll"* (NTSB, 1995a, p. 79). This was due in large part to a lack of experience with three-engine takeoffs. The flight engineer's poor technique when advancing the throttles was therefore classified as a ***skill-based error***.

While the rate at which the throttles were increased was critical to this accident, perhaps a better question is, "Why was the flight engineer at the throttles in the first place?" Recall that this was not the crew's first attempt at

---

[1] To eliminate redundancy, only the final causal category will be presented throughout the remainder of this chapter.

a three-engine takeoff that evening. In fact, after the first aborted attempt, the flight engineer suggested to the captain that "...*if you want to try it again I can try addin' the power...*" (NTSB, 1995a, p. 10), to which the captain agreed. However, the company's operating manual quite clearly states that only the captain, not any other crewmember, is to "*smoothly advance power on the asymmetrical engine during the acceleration to Vmcg*" (NTSB, 1995a, p. 45). In fact, by permitting someone other than the flying pilot to apply asymmetric thrust, the controls became isolated between two individuals with little, or no, opportunity for feedback. In a sense, it was like having someone press on the accelerator of the family car while you steer – certainly not a very good idea in a car, much less a DC-8 with only three operable engines. Given that company procedures prohibit anyone other than the captain to control the throttles during three-engine takeoffs, "*[the crews'] decision to modify those procedures*" (NTSB, 1995a, p. 79) and allow the flight engineer to apply power is considered a violation using HFACS. Furthermore, such violations are arguably rare, and not condoned by management. As a result, this particular unsafe act was further classified as an *exceptional violation*.

Still, this was not the crew's first attempt at a three-engine takeoff that evening. In fact, on their initial attempt, the captain was also unable to maintain directional control because he too had applied power to the asymmetrical engine too quickly. One would think that after losing control on the first takeoff that the conversation during the taxi back into position for the next attempt would largely center on the correct takeoff parameters and three-engine procedures. Instead, more confusion and a continued misunderstanding of three-engine Vmcg and rudder authority ensued, resulting in a second uncoordinated takeoff attempt. In effect, the confusion and "*lack of understanding of the three-engine takeoff procedures*" (NTSB, 1995a, p. 79) exhibited by the crew moments before, and after, the first aborted takeoff, contributed to a general lack of coordination in the cockpit. Consequently, this causal factor was classified as a failure of *crew resource management*.

While it is easy to understand how the confusion with three-engine takeoffs contributed to the accident, it may have been a much less obvious and seemingly insignificant error that ultimately sealed the crew's fate. Even before taxiing out to the runway, during preflight calculations, "*the flight engineer [miscalculated] the Vmcg speed, resulting in a value that was 9 knots too low*" (NTSB, 1995a, p. 77). It appears that instead of using temperature in degrees Celsius, the flight engineer used degrees Fahrenheit when making his preflight performance calculations – an error that is not particularly surprising given that most of the company's performance charts use degrees Fahrenheit, rather than Celsius. Nevertheless, this understandably small error yielded a Vmcg value some nine knots below what was actually required, and may have contributed to the crew's early application of takeoff power to the asymmetrical engine. So yet another error was committed by the crew. However, unlike the conscious decision errors described above, this error was likely one of habit – automatized behavior if you will. After all, the flight engineer was accustomed to using degrees Fahrenheit, not Celsius when making his preflight calculations. It was therefore classified as a *skill-based error*.[2]

It is always easy to point the finger at the aircrew and rattle off a number of unsafe acts that were committed, but the larger, and perhaps more difficult question to answer is, "Why did the errors occur in the first place?" At least part of the answer may lie in the mental state of the crew at the time of the accident. For instance, it was evident from the cockpit voice recorder (CVR) that the crew *"was operating under self-induced pressure to make a landing curfew at [Chicopee]"* (NTSB, 1995a, p. 77). Comments by the captain like the need to get *"as much direct as we can"* (referring to the need to fly as direct as possible to Chicopee) and *"...we got two hours to make it..."* as well as those by the first officer such as *"...boy it's gettin' tight..."* and *"...[hey] we did our best..."* all indicate a keen awareness of the impending curfew and only served to exacerbate the stress of the situation (NTSB, 1995a, pp. 3, 14, and 97). Although we will never really know for certain, it is conceivable that this self-induced time pressure likely influenced the crew's decisions during those critical moments before the accident – particularly those of the captain when deciding to continue rather than abort the takeoff. Using the HFACS framework, we classified the self-induced pressure of the aircrew as a precondition for unsafe acts, specifically an *adverse mental state*.

In addition to self-induced pressure, it appears that the crew, especially the captain, may have suffered from mental fatigue as well. As noted earlier, during the previous two days, the crew had flown a very demanding schedule. Specifically, they had flown a 6.5-hour check ride from Dover, Delaware to Ramstein, Germany, followed by a little less than 10 hours rest. This was followed by a 9.5-hour flight from Ramstein to Dover via Gander, Newfoundland. Then, after arriving in Dover, the crew checked into a hotel at 0240 for what was to have been 16 hours of crew rest before they could legally be assigned another commercial (revenue generating) flight.[3]

---

[2] The inconsistencies among the performance charts could be considered causal as well. However, because it was not cited as such by the NTSB, we will not pursue that issue here.

[3] The rules and regulations associated with commercial and general aviation can be found in the Code of Federal Regulations Volume 14 (Aeronautics and Space) under 14 CFR Part 91 (general aviation), 14 CFR Part 119 (air carrier and commercial operations), and 14 CFR Part 121 (domestic, flag, and supplemental operations).

However, rather than get a good night's sleep, the captain's scheduled rest was disrupted several times throughout the night by the company's operations department and chief pilot to discuss the additional ferry flight the next day. In fact, telephone records indicate that the captain's longest opportunity for uninterrupted rest was less than five hours.

So after only about 12 hours of rest (but not necessarily sleep) the captain and crew departed Dover for their first flight of the day, arriving in Kansas City shortly after 5:30 pm. Unfortunately, fatigue is often a nebulous and difficult state to verify, particularly when determining the extent to which it contributes to an accident. That being said, the poor decisions made by the aircrew suggest that at a minimum, the captain, and perhaps the entire crew, *"were suffering from fatigue as a result of the limited opportunities for rest, disruption to their circadian rhythms, and lack of sleep in the days before the accident"* (NTSB, 1995a, p. 76), all resulting in an ***adverse mental state*** using the HFACS framework.

Up to this point we have focused solely on the aircrew. But, what makes HFACS particularly useful in accident investigation is that it provides for the identification of causal factors higher in the system, at the supervisory and organizational levels. For instance, one might question why this particular aircrew was selected for the ferry flight in the first place; especially when you take into consideration that other, more experienced crews, were available. As you may recall, there was no record of the captain having previously performed a three-engine takeoff as pilot-in-command, nor had any of the crewmembers even assisted in one. Even so, company policy permitted all of their DC-8 captains to perform this procedure regardless of experience or training. As a result, the crew was "qualified" for the flight by company standards; but were they by industry standards? In fact, of the nine other cargo operators contacted by the NTSB after this accident, only two used line flight crews (like the accident crew) for three-engine takeoffs. Instead, most cargo operators used check airmen or special maintenance ferry crews exclusively for these flights. Of the two that did not, one restricted three-engine takeoffs to only the most experienced crews. Regardless of the rationale, the assignment of this crew to the three-engine ferry flight when *"another, more experienced, flight crew was available"* (NTSB, 1995a, p. 77) was considered an instance of unsafe supervision, in particular, ***planned inappropriate operations***.

So, why did the company choose not to assign its best crew to the three-engine ferry flight? Perhaps the company's decision was driven in part by current U.S. Federal Air Regulations that require pilots flying revenue generating operations (i.e., carrying passengers or cargo) to get at least 16 hours of rest before flying another revenue flight, but places no such restrictions on pilots conducting non-revenue flights. As a result, the accident crew was permitted to fly the non-revenue ferry flight from Dover to Kansas City with only about 9.5 hours of rest and could then fly the non-revenue, three-engine ferry flight to Chicopee. Because of the Federal regulations however, they could not legally fly the revenue generating flight scheduled to fly from Kansas City to Toledo that same day. In fact, another crew in the area had more experience flying three-engine ferry flights, but they were also eligible under Federal guidelines to fly revenue-generating flights. So rather than use a crew with more experience in three-engine ferry flights for the trip to Chicopee and delay the revenue flight to Toledo, the company chose to use a crew with less experience and get the revenue flight out on time. It can be argued then, that the decision to assign the ferry flight to the accident crew was influenced more by monetary concerns than safety within the company. Within the HFACS framework, such organizational influences are best considered breakdowns in *resource management*.

Remarkably, the FAA, due to a previous DC-8 crash, had known about the hazards of flight and duty time regulations that *"permitted a substantially reduced crew rest period when conducting [a] non-revenue ferry flight [under 14 CFR Part 91]"* (NTSB, 1995a, p. 77). In fact, the NTSB had previously recommended that the FAA make appropriate changes to the regulations, but they had not yet been implemented. So, where do we capture causal factors external to the organization within the HFACS framework? You may recall from earlier chapters that causal factors such as these are not captured within the HFACS framework per se, because they are typically not within an organization's sphere of influence (i.e., the organization has little or no control over Federal rulemaking). Instead, they are considered *outside influences* (in this case, loopholes within Federal Air Regulations) that have the potential to contribute to an accident.

Still, some may argue that the crew had been trained for three-engine ferry flights, the same training that the company had provided to all their captains. As such, they were "qualified" to fly the mission. But was that training sufficient? The knee-jerk reaction is obviously 'no' given the tragic outcome. But some digging by the NTSB revealed an even larger problem. It appears that the simulator used by the company did not properly simulate the yaw effect during a three-engine takeoff – the very problem the crew experienced during both takeoff attempts. Indeed, a post-accident test of the simulator revealed that with only three of the four engines brought up to takeoff power, the runway centerline could be easily maintained regardless of the airspeed achieved, something that clearly cannot be done in the actual aircraft. Regrettably, this lack of realism went unnoticed by the company's training department. As a result, the lack of *"adequate, realistic training in three-engine takeoff techniques or procedures"* (NTSB, 1995a, p. 76) within the company training program was viewed as yet another precondition for unsafe acts (environmental factor), in particular a failure within the ***technological environment***.[4]

In addition to the limitations inherent within the simulator itself, *the three-engine takeoff procedure description in the airplane operating manual was confusing* (NTSB, 1995a, p. 76). Unfortunately, the dynamics of a three-engine takeoff are such that even the proper application of asymmetric throttles leaves little margin for error. Any confusion at all with the three-engine procedures would only exacerbate an already difficult task. To their credit however, the company did provide regular training of the three-engine takeoff procedure, but the poor description of the maneuver in the operations manual, and inaccurate simulator portrayal, lessened its effectiveness. Therefore, we did not consider this an organizational process issue. Instead, the procedural shortcomings in the company's operating manual, like the trouble with the simulator, were considered a failure of the ***technological environment***.

Finally, the lack of government oversight on occasion will affect the conduct of operations. That was once again the case in this accident as the NTSB found that the FAA was not effectively monitoring the company's domestic crew training and international operations. A post-mishap interview with the FAA's principal operations inspector (POI) for the company revealed a general lack of familiarity with their CRM training program, crew pairing policies, and several aspects of the company's ground training program at Denver. This lack of oversight was likely the result of manning cuts and funding issues that restricted the POI's ability to travel to Denver from his base in Little Rock, Arkansas. The lack of government oversight is yet another example of the type of ***outside influence*** that can affect operations and, like the issues regarding Federal regulations guiding crew rest above, is included here for completeness.

Summary

As with most aviation accidents, indeed most accidents in general, this one could have been prevented at many levels (Figure 4.3). While the lack of governmental oversight of the company's operations and the loopholes contained within the FAA's crew rest guidelines may have laid the foundation for the tragic sequence of events to follow, the failures within the organization itself were far more pervasive. In a sense, the crew was "set-up" by a lack of realistic training (recall the problems with the simulator), a

---

[4] Note that this particular technological failure did not occur in the actual aircraft as is typically seen in aviation accidents. Instead, the failure was in the simulator that this, and other crews trained on, thereby contributing to the tragic sequence of events that followed.

confusing description of three-engine takeoffs in the manual, and perhaps most important, poor crew manning practices.

Nevertheless, the crew was also responsible. Clearly the decisions made in the cockpit that night were marred by the failure to thoroughly understand the intricacies of three-engine takeoffs which was exacerbated by poor crew coordination, mental fatigue, and a sense of urgency to get underway. Even so, had the captain aborted the takeoff, as he had done previously when the aircraft lost directional control, the accident would have been averted. Instead, he chose to continue the takeoff by initiating rotation below computed rotation speed, leading to a further loss of control and collision with the terrain.

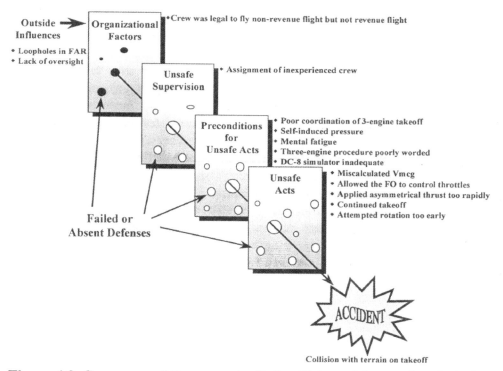

Figure 4.3 **Summary of the uncontrolled collision with terrain of a DC-8 at Kansas City International Airport**

A World Cup Soccer Game They would Never See

Shortly before midnight, in June of 1994, a Learjet 25D operated by Transportes Aereos Ejecutivos, S. A. (TAESA) departed Mexico City with 10 passengers bound for a World Cup Soccer game in Washington, DC. Flying through the night, the chartered flight contacted Washington Dulles Approach Control just before six o'clock in the morning and was informed that conditions at the field were calm with a low ceiling and restricted visibility due to fog. Moments later, using the instrument landing system (ILS), the crew made their first landing attempt. Having difficulty maintaining a proper glideslope, the crew was forced to fly a missed approach and try again. Sadly, on their second attempt, they again had difficulty maintaining a proper glideslope, only this time they fell below the prescribed glide path and impacted the trees short of the runway, fatally injuring everyone on board (NTSB, 1995b).

The crew's preparation for the flight began normally as both crewmembers reported for duty around 10:00 pm appearing well rested and in good spirits. In fact, according to his wife, the captain had been off duty for three days and even took a 3-hour nap before reporting that evening. There was no reason to believe that the first officer was fatigued either, although it remains unclear how much sleep he actually got since there were no witnesses.

With the necessary preflight activities completed and the passengers loaded, the crew departed Mexico City shortly after 11:00 pm for what was to be a chartered flight to Washington, DC. After pausing briefly for a planned refueling stop in New Orleans, the flight continued uneventfully until they entered the Washington, DC area. There, after a brief delay for an aircraft with a declared emergency, the crew contacted Dulles Approach Control and was advised that due to low visibility (approximately 1/2 mile) and ground fog at the field, a straight-in, ILS-coupled[5] approach (referred to as a Category III approach) would be required.

---

[5] The ILS is a radio-navigation system designed to help pilots align their planes with the center of the runway during final approach under conditions of poor visibility. The system provides final approach glide path information to landing aircraft, allowing pilots to land safely. Using two directional transmitters on either side of the runway centerline, the ILS provides pilots with horizontal (referred to as stabilizer or localizer) and vertical (referred to as glideslope) guidance just before, and during, landing. Typically, this information is shown on an instrument display in the form of horizontal and vertical lines, which enables the pilots to determine their exact position in relation to the runway and maneuver their aircraft accordingly.

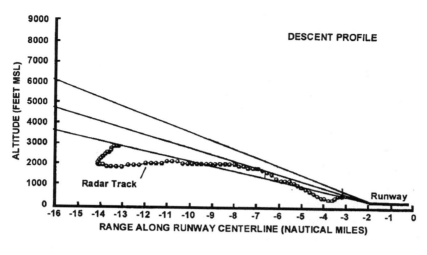
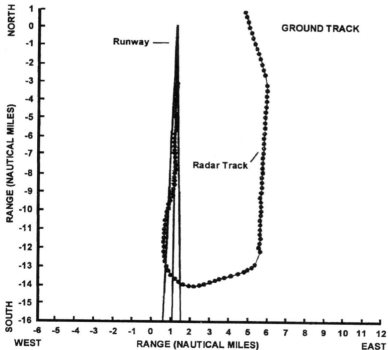

Figure 4.4 Aircraft descent profile and ground track during the accident approach
Source: NTSB (1995b)

So, with the onset of daybreak, the crew began their first ILS-coupled approach to runway 19R. The approach started out well as the aircraft remained within prescribed localizer parameters (i.e., they had not deviated too far left or right of their assigned path) approximately 14 nautical miles (nm) from the runway threshold. However, the aircraft was never fully stabilized on the glideslope, flying erratically (up and down) along the approach path. Evidence from the Air Traffic Control radar track had the aircraft briefly leveling out at 600 feet for a few seconds, perhaps in an attempt to establish visual contact with the airport – something that, if true, was clearly not authorized. In fact, with the runway obscured by fog, company procedures and U.S. Federal regulations required that the captain abort the landing and follow the published go-around procedures for another attempt or abandon the landing altogether. Nevertheless, it was only after the air traffic controller inquired about their intentions that the crew broke off their approach and accepted vectors for another try.

The second approach was initially more stable than the first as can be seen from Figure 4.4. However, it quickly resembled the first, and from the outer marker (roughly 6.5 miles from the runway threshold) to time of impact, the aircraft flew consistently below the published glideslope. Eventually the aircraft impacted the trees and crashed a little less than one mile short of the runway threshold.

Human Factors Analysis using HFACS

As with the previous case study we will work backwards from the aircraft's impact with the terrain. In this case, what ultimately led to the accident was the captain's *"failure to adhere to acceptable standards of airmanship during two unstabilized approaches"* (NTSB, 1995b, p. 46). What we do know from the factual record is that during both approaches to Dulles, the TAESA aircraft flew well below the published glideslope. However, what remains unclear is whether or not the captain intended to go below the glideslope in an effort to establish visual contact with the airport while flying in a thick fog. As described above, there appears to be some evidence from the radar track that the pilot tried to obtain a visual fix on the runway during the first approach.

Regrettably, the practice of flying below cloud layers and other adverse weather in an attempt to get a view of the ground is not altogether uncommon in aviation, just extremely dangerous and inconsistent with published Category III procedures. However, unlike the first approach, there appears to be no clear indication that the pilot intended to go below the glideslope on the second attempt. Actually, it is probably more plausible that the captain simply did not possess the necessary skill to safely perform an instrument approach. As we will see later in the analysis, the captain was not particularly adept at these types of approaches. This, coupled with his possible desire to gain visual contact with the ground, may have led to a breakdown in instrument scan and quite possibly explains why the captain was unaware that he was fatally below the glideslope. In the end, one thing is certain, the crew clearly did not know how dangerously close to the terrain they were; something an active instrument scan would have told them. Therefore, the failure of the aircrew to properly fly the ILS approach can be largely attributed to a breakdown in instrument scan and the inability of the pilot to fly in instrument conditions, in both cases a *skill-based error* using HFACS.

With the weather conditions at Dulles rapidly deteriorating, one could question why the second approach was even attempted in the first place, particularly given that category III operations were in effect. These operations require that aircraft fly a straight-in, ILS-coupled, approach to reduced visual minimums and require special certifications of the crew, runway, and equipment. In this case, however, the crew was only authorized to perform a category I approach that establishes a 200 foot ceiling and at least ½ mile (2400 feet) visibility at the runway for an approach to be attempted. Often referred to as runway visual range (RVR), the visibility at the runway had deteriorated well below the 2400-foot minimum at the time of the accident. In fact, just moments before impact the local controller confirmed that the aircraft was on the ILS and advised them that RVR at touchdown and at the midpoint of the runway was 600 feet. Nevertheless, the captain continued the approach; one that *"he was not authorized to attempt"* (NTSB, 1995b, p. 46); a clear violation of the Federal regulations in effect at the time of the accident.

While we can be certain that a violation was committed, determining what type (routine or exceptional) is much less clear. In reality, making this determination is often difficult using NTSB or other investigative reports, primarily because many investigators do not ask the right questions nor do they dig deep enough to reliably classify the type of violation. As a result, we are usually left only with the observation that a violation was committed. So, without supporting evidence to the contrary, we have not classified this particular violation as either routine or exceptional; instead, we simply stopped with the overarching category of *violation*.[6]

But, why would the crew attempt an approach beyond their legal limits – particularly given the first unsuccessful ILS attempt? After all, it is not as if the crew had no options. For instance, they could have waited for the weather to improve. Or perhaps they could have elected to switch runways from 19R

---

[6] Unfortunately, many investigations within military and civilian aviation have not investigated their accidents sufficiently to allow a reliable classification of routine, versus exceptional violations. Recently however, the U.S. Navy/Marine Corps have incorporated this distinction in their aviation accident investigations to improve the quality of their investigations.

to 19L where visibility was noticeably better. In fact, another commercial flight was offered that option earlier, but elected to divert to Pittsburgh instead. Indeed, that may have been the most prudent alternative. The TAESA crew could simply have diverted to their designated alternate (Baltimore International Airport), about a hundred miles northeast of Dulles. Unfortunately, rather than "*[hold] for improvements in weather, [request] a different runway (19L), or [proceed] to the designated alternate airfield*" (NTSB, 1995b, p. 46), the crew elected to continue the approach – perhaps in the hope that a window of visibility would present itself. Such a decision, in light of safer alternatives, was classified a *decision error* using HFACS.

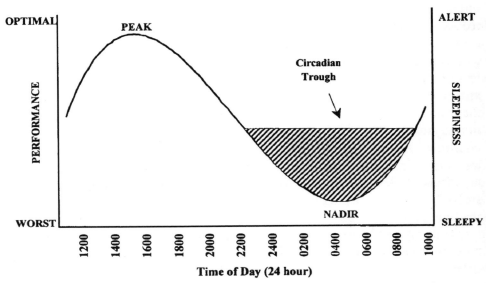

Figure 4.5 Sleepiness and performance as a function of time of day

While it is easy to identify what the crew did wrong, to truly understand how this tragedy could happen, one needs to explore the preconditions that existed at the time of the accident. First, and perhaps foremost, "*the crew may have been experiencing the effects of fatigue following an all-night flight*" (NTSB, 1995b, p. 47). It is a well-known scientific fact that performance degrades and errors, particularly decision errors, increase between 10:00 pm and 06:00 am for individuals entrained to the normal light-dark cycle (Figure 4.5). Otherwise known as the circadian trough, the majority of the flight took place during this time. In fact, when the accident occurred (just after 04:00 am Mexico City time), the crew was likely performing during a low point (nadir) in the circadian rhythm and experiencing the adverse effects of mental fatigue. When this is coupled with the fact that the crew had been flying for several hours with only a short stop in New Orleans for fuel, it is reasonable to assume that they were tired, an *adverse mental state* using the HFACS framework.

But fatigue was not the only thing that affected the decisions of the crew that morning. Also contributing to the accident was the "*relative inexperience of the captain for an approach under these conditions*" (NTSB, 1995b, p. 46). Among major air carriers in the U.S., upgrades to captain usually do not occur until the pilot has amassed somewhere between 4,000 and 5,000 flight hours. In some countries however, where the number of qualified pilots may be limited, it is not at all unusual for candidates to have around 2,300 hours when upgrading. Even so, this captain had actually less experience than that, having accumulated just over 1,700 total flight hours, of which less than 100 were as pilot-in-command (PIC). Compounding the problem, it had been noted during the captain's recurrent training that his crew management and decision-making skills, while adequate under normal conditions, needed improvement when he was placed under stress or in emergency situations. As a result, it was recommended that he fly with a strong training captain or first officer during his upgrade. Unfortunately, he could not rely on the first officer either since he was also relatively inexperienced, having accumulated just over 850 total flying hours, only half of which were in the Learjet. Ultimately then, it can be argued that the captain's lack of experience as PIC, as well as his difficulties under stressful conditions like those that prevailed at the airfield, limited his ability to make sound decisions, and was therefore considered a *physical/mental limitation*.

This begs the question, "How could a pilot with such limitations be upgraded to captain?" The answer, it seems, resides with the "*ineffective communications between TAESA and ... the contract training facility regarding the pilot's skills*" (NTSB, 1995b, p. 46). The official evaluation provided to TAESA by the training facility described the captain as "*focused*" and "*serious*," a "*smooth pilot*" and a "*polished first officer*." This description was in stark contrast to what was actually printed in the instructor's notes that described him as a pilot with only satisfactory flying skills during normal flying conditions and documented problems with his instrument scan and crew coordination. Furthermore, it was noted that "*although he flew non-precision approaches well ... his instrument approaches definitely did not meet ATP [airline transport pilot] standards*" (NTSB, 1995b, p. 7). Unfortunately, only the "official" evaluation reached TAESA. As a result, permissive language contained in the report opened the door for TAESA officials to interpret the document as approval of their applicant. Had the instructor's notes, which painted a clearer picture of the captain's below average performance, been made available to TAESA, or had the evaluation been worded more clearly, the training facilities intent may have been more plainly communicated and a delay in this pilot's upgrade to captain might have occurred. In fact, it was only after repeated requests by the Director of Operations that a letter was finally received describing the captain's need *"to improve his airmanship and command skills"* as well as the need for *"situational awareness training under high workload"* (NTSB, 1995b, p. 8). Clearly, there was a miscommunication between TAESA and the contract facility, which led to the decision to upgrade a captain with limited experience. Recognizing that both organizations were responsible for the miscommunication (on an organizational level), we classified this causal factor as a failure of *resource management*.

That being said, *"oversight of the accident flight by TAESA was [also] inadequate."* It turns out that the operations specifications in use by TAESA at the time of the accident failed to address which visibility value, RVR or prevailing, takes precedence in establishing a minimum for landing. Had this been a simple oversight of one particular aspect of the operational specifications, we might have chosen to classify this as a failure within the technological environment (i.e., inadequate documentation). However, a review of TAESA's operation specifications revealed that some of the pages were as much as 20 years old and none addressed the precedence of RVR or prevailing visibility on landing. We therefore classified the failure of TAESA to address this provision within the company's operations manual constitutes as a failure of *operational processes*.

Finally, the NTSB suggested that an operating ground proximity warning system (GPWS) might have prevented the accident. Although the failure to equip the aircraft with a functional GPWS is not a violation per se, it was considered a factor in this accident. TAESA was operating under 14 CFR, part 129 of the U.S. Federal Code of Regulations, which did not require installation of a GPWS for this aircraft. However, a GPWS aboard the aircraft likely would have provided a continuous warning to the crew for the last 64 seconds of the flight and may have prevented the accident. Given that controlled flight into terrain continues to plague aviation, a GPWS seems a reasonable fix in many instances. Although not a violation of any established rule or regulation, operating aircraft without such safety devices is typically a decision made at the highest levels of an organization. We therefore classified this factor as a failure of *resource management*.

Summary

As with most accidents, this accident unfolded with errors at several levels (Figure 4.6). At the company level, a decision to install a ground proximity warning system in all turbojet aircraft may have prevented this accident. The company leadership also showed poor planning by placing a very inexperienced captain with an even less experienced first officer. Not only was it questionable that the captain was even certified at the PIC level; but, if he was required to fly as a captain, he should have been paired with a very experienced First Officer to offset his lack of experience. Wisely, the captain took a nap the afternoon before the departure at Mexico City. Unfortunately, one nap may not have been enough to overcome the increase in fatigue and performance degradation associated with the circadian trough. In the final analysis, these breakdowns certainly contributed to the captain's decision to continue the ILS approach, one that he was incapable of flying safely that morning.

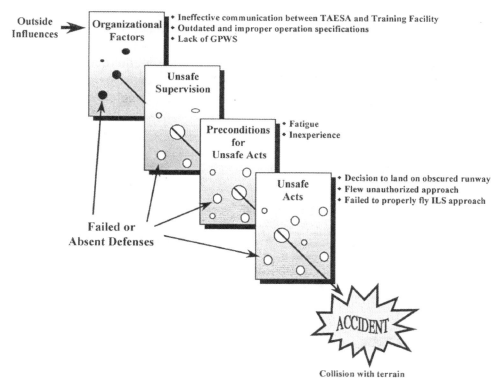

Figure 4.6 Summary of the controlled flight into terrain of the Learjet one mile short of Dulles International Airport

The Volcano Special

On the afternoon of April 22, 1992, a Beech Model E18S, took off from Hilo, Hawaii with a group of tourists for the last leg of a sight-seeing tour of the Hawaiian Islands chain. Marketed as the

"Volcano Special," the aircraft was to fly over the Islands of Molokini, Lanai, and Molokai on the western edge of the Hawaiian Island Chain. However, this day the young pilot inexplicably flew off course and into a layer of haze and clouds that obscured much of the Island of Maui and Mount Haleakala that lay just North of his planned route. Moments later, the aircraft crashed into the side of the Volcano, killing all eight tourists and the pilot (NTSB, 1993).

The day began innocently enough, as eight tourists boarded Scenic Air Tours (SAT) Flight 22 for a day of sightseeing along the Hawaiian Island chain. The first part of the tour included sights along the north coast of Molokai and Maui, culminating with a flight over the Kilauea Volcano on the "Big Island" of Hawaii. After landing at Hilo Airport shortly after nine in the morning, the passengers were shuttled off for a 6-hour ground tour of sights around the Island of Hawaii. Later that afternoon, the tourists returned to the airport and boarded SAT Flight 22 for a 3:20 departure to Honolulu via the western edge of the Hawaiian Island chain. Tragically, they never made it past Maui.

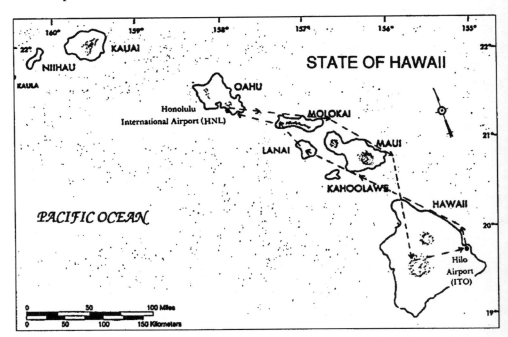

Figure 4.7 Planned tour route of SAT flights
Source: NTSB (1993)

Unfortunately for the tourists, the weather that afternoon was not particularly favorable to sightseeing over the interior of the islands as haze and rain showers had settled in, lowering the visibility to three miles in some areas. Still, the weather, in and of itself, did not present a problem for the pilot as the normal flight path for SAT flights returning to Honolulu did not include flying over the interior of the islands. Instead, the typical flight to Honolulu included sights along the northern coast of Hawaii to Upola Point, at which time the aircraft would fly northwest over the village of Makena on the southern shore of Maui (Figure 4.7). Passing over Lanai, the flight would then normally continue to Honolulu and land at the airport.

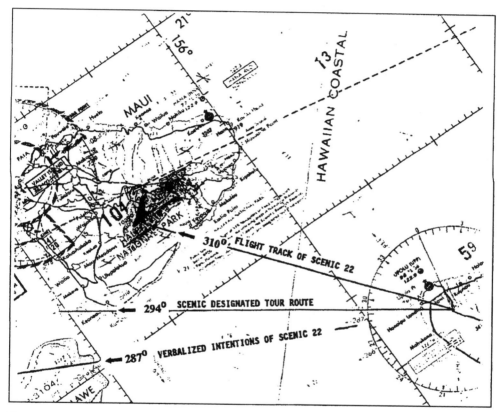

Figure 4.8 Designated, planned, and actual flight path of SAT Flight 22
Source: NTSB (1993)

The return flight that day however, was anything but normal. Shortly after takeoff, the pilot of Flight 22 called the Honolulu Flight Service Station to inquire about the status of a restricted area over the island of Kahoolawe, just south of the company's normal route. When advised that the area was closed between the surface and 5000 ft mean sea level (msl), he advised air traffic control (ATC) that he would be flying over the top at 6500 feet. As can be seen from Figure 4.8, this slight deviation from his original flight plan would require that he fly a magnetic compass heading of 287° from the Upola Point VOR (very high omnidirectional radio range), located on the northwest point of the island of Hawaii.

The normal route flown by company pilots would typically have them flying a heading of 294° from Upola Point along the western edge of Maui. However, this day, the pilot of SAT Flight 22 flew neither the 294° nor the 287° heading. Instead, he inexplicably flew a 310° heading directly into the interior of Maui and multiple layers of clouds that had formed, obscuring Mount Haleakala from view.

Passing the shoreline of Maui at roughly 8100 ft., SAT Flight 22 continued to ascend until its last recorded altitude of 8500 ft., just prior to impacting the volcano. A witness in the area at the time of the accident reported that while he did not see the impact due to heavy, rolling clouds in the area, he did hear what he believed to be a multi-engine plane until the sound stopped abruptly. Little did he know, the silence was the result of a tragic plane crash that took the lives of the pilot and his eight passengers.

Human Factors Analysis using HFACS

Working backwards from the impact, it does appear that in the final seconds before the accident that the captain did try to avoid the volcano. Regrettably however, he did not see the rising terrain of Mount Haleakala until the final seconds of flight because the cloud cover obscured it. In fact, by the time he realized the tragic error he had made, his fate and that of the passengers was likely already sealed.

But why did Flight 22 enter the clouds in the first place? Perhaps he thought that he would simply pass harmlessly through the clouds and continue the tour on the other side. But any flight into the clouds is seemingly beyond comprehension when you consider that this was a "sight-seeing" flight and passengers certainly cannot see much in the clouds. What's more, SAT was a VFR-only operation. In other words, according to FAA regulations, all flight operations were required to be flown in visual meteorological conditions (VMC) while under visual flight rules (VFR), thereby prohibiting flight into the clouds and weather. Although we will never know exactly why the captain chose to fly into the clouds, what we do know is that contrary to regulations that required SAT flights *"to be conducted under VFR, the captain chose to continue visual flight into the instrument meteorological conditions that prevailed along the eastern and southern slope of Mount Haleakala on the Island of Maui"* (NTSB, 1993, p. 46) – a clear *violation* of existing company and FAA regulations. What is less clear is what type of violation was committed. Therefore, as with the preceding case study, we chose not to classify this causal factor further.

It can only be assumed that the captain was aware of the rules regarding staying clear of weather and the clouds and the dangers associated with disregarding them. This only makes it more difficult to understand why he would elect to fly into the clouds rather than circumnavigate them. One possibility is that he did not see the clouds until after he was in them. This seems a bit improbable since he was told by personnel at the local Flight Service Station that VFR-flight was not recommended along the interior of the islands because of haze and moderate rain showers. What is more likely is that he did not realize that the upsloping cloud layer in front of him was produced by Mount Haleakala. Even this hypothesis is a bit surprising since pilots knowledgeable of weather patterns in the Hawaiian Islands would realize that only a landmass like Mount Haleakala could generate the orographic lifting of clouds at the altitudes that Flight 22 encountered them. Perhaps this is part of the problem since, as we shall see, this particular pilot had limited experience flying among the Hawaiian Islands. As a result, we chose to classify the fact that the "...*captain did not evaluate the significance of an upsloping cloud layer that was produced by [the] orographic lifting phenomenon of Mount Haleakala*" (NTSB, 1993, p. 46) as a knowledge-based *decision error*.

Still, the accident might not have occurred at all had the captain not deviated from his intended course of 287°. Recall that he reported to the Honolulu Flight Service Station that he would be flying over Kahoolawe, considerably south of the 310° course that he actually flew. How then, could such a tragic error be committed? The NTSB considered many plausible reasons including the possibility that the captain might have wanted to show the Mount Haleakala volcano crater to the passengers. However, because of existing weather and schedule considerations, the NTSB ultimately ruled out any intentional deviation. So, what could have led to such a tragic error? While we will never really know, perhaps the most plausible reason may be the failure of the captain to refer to aeronautical references for navigation information. Indeed, three Hawaiian island VFR sectional charts were found folded among the wreckage inside the captain's flight bag. It would appear then that the captain "*did not use his navigation charts to confirm the correct heading and radial outbound from Upolu Point*" (NTSB, 1993, p. 47). By not using all available information at his disposal, the captain did not practice good *crew resource management.*

But, this still does not explain how the aircraft ended up on the 310°, vice 287° radial. One explanation lies with the omni-bearing selector, or OBS as it is called. The OBS is a dial used to select the desired radial one would like to fly on, in this case the 287° radial. Given the aircraft's flight path, it is quite possible that the captain failed to turn the omni-bearing selector (OBS) to the 287° radial while tracking outbound from the Upola VOR (Figure 4.8).

Curiously, the bearing the aircraft actually flew (310°) is identical to the radial SAT pilots typically fly when outbound from Hilo Airport in order to fly past some popular attractions along the north shore of Hawaii. In fact, the same 310° radial is the initial flight track for the northern route to Honolulu via Hana, a route that the captain had flown *four* times in the five days prior to the accident. It seems entirely plausible that after tuning in the Upola VOR, that the captain either simply forgot to select the 287° radial using the OBS or selected the 310° radial out of habit. In either case, a *skill-based error* was committed. Furthermore, his *"navigation error went undetected because he failed to adequately cross-check [his] progress ... using navigation aids available to him"* (NTSB, 1993, p. 47). This seemingly simple breakdown in instrument scan is yet another *skill-based error* committed by the captain.

As with the other two case studies, there was more to this tragic story as well. It turns out that the captain was a former van driver with SAT – not once, but twice. In fact, he had been employed by nine different employers (twice with SAT) in the five years before the accident. Why this is important is that five of those employers had fired him because of *"misrepresentation of qualifications and experience, failure to report for duty, disciplinary action, poor training performance, and work performance that was below standards"* (NTSB, 1993, p. 14). This is particularly relevant here, because the captain had once again misrepresented his credentials when applying for the pilot job with SAT in the summer of 1991. Upon employment, the captain indicated that he had roughly 3,400 flight hours of which 3,200 were as PIC including 1,450 hours in twin-engine aircraft and roughly 400 hours of instrument time. However, using FAA records, he actually had fewer than 1,600 hours and less than 400 of those were in multiengine aircraft. Even the most liberal account had the captain with less than 2,100 hours, still well short of the 2,500 hours of actual flight time (including 1,000 hours of multiengine experience) required by SAT for employment. In fact, he had not met those requirements by the time the accident occurred! Clearly, this deceitful act contributed to the captain's ability to safely fly tourists around Hawaii; but, where do you classify the fact that *"the captain falsified [his] ... employment application and resume when he applied for a pilot position at Scenic Air Tours"* (NTSB, 1995b, p. 46) within the HFACS framework? Clearly, it was not an unsafe act violation, because it did not have immediate consequences, nor did it happen in the cockpit. What it did do was did set the stage for the unsafe acts to follow and affected the captain's overall readiness to perform. Consequently, it was considered a failure of *personal readiness* on the part of the pilot.

Perhaps even more difficult to understand is how a pilot with an employment and flight history as checkered as the captain's could be hired in the first place. After all, didn't SAT conduct a preemployment background check of the captain's employment and aeronautical experience? Recall that even at the time of the accident *"the pilot did not possess the minimum hours of experience stipulated in the company operations manual to qualify as a captain"* (NTSB, 1993, p. 46). Indeed, several potential employers had either fired or refused to hire him when it was determined that his experience did not meet their standards and at least one airline rejected the captain's application when it was determined that he had misrepresented his credentials. The fact that *"SAT was unaware of the captain's falsified employment application because they did not [conduct a] ... substantive preemployment background check"* (NTSB, 1993, p. 46) is therefore considered a classic failure of human *resource management*.

But how could this happen? Aren't operations like SAT's required to conduct a standard background check? Actually, at the time of the accident, *"the FAA did not require commercial operators to conduct substantive pilot preemployment screening"* (NTSB, 1993, p. 47). This was somewhat surprising since the NTSB had previously identified problems with preemployment screening, recommending that commercial operators be required to conduct substantive background checks of pilot applicants. Unfortunately, while the FAA agreed with the intent of the recommendations, it did not believe that the benefits would outweigh the costs of promulgating and enforcing them (NTSB, 1993, p. 39). Consequently, the intentional inaction by the FAA is considered an *outside influence*.

Summary

As with the other two case studies, this tragic accident involved more than the unsafe acts of the pilot alone. While there is no denying that the decision to fly into IMC and the navigational errors were attributable to the captain, the fact that he was flying passengers in that situation was the responsibility of SAT. After all, had the company done an adequate background check, as at least one other airline had, they would have realized that the captain had misrepresented his experience and likely would have uncovered an employment history that might have influenced their decision to hire him. Certainly, the FAA bears some responsibility as well, since there were no regulations in place that required commercial operators to conduct substantive pilot **preemployment** screening. The good news is that this has since been remedied by the FAA.[7]

---

[7] With the passage of the Pilot Records Improvement Act in 1996, all air carriers are now required to perform a background check of all pilot applicants before allowing an individual to begin service. Up-to-date information can be found in FAA Advisory Circular 120-68B.

98 *A Human Error Approach to Aviation Accident Analysis*

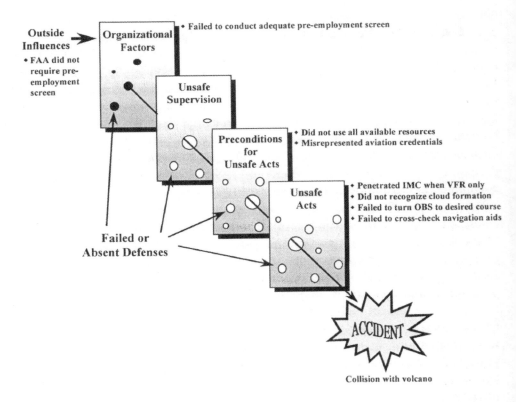

Figure 4.9 Summary of the in-flight collision with Mount Haleakala, Maui, Hawaii

Conclusion

The preceding case studies demonstrate how HFACS can be used to classify existing human causal factors contained within NTSB and other accident reports. In fact, we have used similar civilian and military cases within our own workshops and training seminars to demonstrate the ease with which HFACS can be used to analyze accident data. Although it is impossible to demonstrate in a book, we hope that the reader can see how a similar process could be used to identify human casual factors in the field during an actual accident investigation. The U.S. Naval Safety Center, for example, has improved the richness of the human error data significantly since HFACS was introduced to the aviation safety community. What remains to be answered however, is what can be done with the data once it is collected. For answers to that and other questions, we will turn to the next chapter.

5 Exposing the Face of Human Error

After spending the weekend visiting with the pilot's parents, the crew of a U.S. Navy F-14 arrived at the hangar for a scheduled flight later that morning. But today, unlike previous missions, the pilot's parents and a family friend accompanied the crew as they busily prepared for the flight. After completing their pre-flight checks and pausing briefly for some pictures, the pilot and his radar intercept officer (RIO) strapped themselves into their multi-million dollar jet in preparation for their flight home. Then, with a sharp salute to his father, the pilot taxied his jet into position and waited as his parents hurried off to proudly watch the F-14 depart. Little did they realize that they would soon witness the tragic loss of their son.

The pilot was a second tour fleet aviator known more for his abilities and performance as a Naval Officer than for his skills as a pilot. Indeed, he had always been a marginal aviator, struggling to prove his worth – especially after an accident less than one year earlier in which he lost control of his aircraft and ejected from an unrecoverable flat spin.

But that morning, as he sat ready to launch in his Navy Tomcat, he was the pride of the family. With his parents watching from a restaurant just beyond the end of the runway, he began his takeoff roll. Reaching nearly 300 knots he rapidly pulled the nose of his jet up, zorching for the cloud layer less than 2,500 feet above. Never before had he taken off with such an extreme pitch attitude nor was he authorized to do so, particularly in poor weather conditions. But this was no ordinary day.

Although he had been cleared by air traffic control for an unrestricted climb just moments earlier, the pilot began to level off after entering the cloud layer – perhaps believing that the altitude restriction of 5,000 feet was still in effect. Unfortunately, by transitioning to level flight in the clouds after a high g-force takeoff, the crew rapidly fell prey to the disorienting effects of the climb. Presumably flying by feel rather than instruments, the pilot continued to drop the nose of his jet until they were roughly 60° nose down and heading straight for the ground a few thousand feet below. Just seconds later, the aircraft exited the cloud layer still some 60–80° nose down with an airspeed in excess of 300 knots. Seeing the world fill his windscreen rather than the blue sky he expected, the pilot abruptly pulled back on the stick and selected afterburner in a futile attempt to avoid hitting the terrain. In the end, he lost control of the aircraft as it crashed into a residential area killing himself, his RIO, and three unsuspecting civilians.

For obvious reasons, impromptu air shows like this one for family and friends are prohibited and arguably rare within the U.S. military. On the other hand, when they do occur they often make front page news and understandably leave American taxpayers questioning the leadership and professionalism of our armed forces. But, are military pilots the unbridled risk takers portrayed in Hollywood movies like *Top Gun* or are they a class of highly educated, elite warriors involved in a high stakes occupation worthy of our praise and admiration? While the latter is surely the case, those at the highest levels of the military are often at a loss when explaining how a responsible Naval aviator could find himself in a no-win situation like the one described above.

Conventional wisdom has historically been our only means for addressing this issue. Even the most senior aviators may only be personally familiar with a handful of accidents, of which few, if any were associated with the willful disregard for the rules. So when asked very pointed questions such as, "Admiral, how many accidents in the U.S. Navy are due to violations of the rules?" they very honestly reply, "Very few, if any." In fact, even the most experienced analysts at premiere safety organizations like the U.S. Naval Safety Center have traditionally had little more to offer. Rather than answer the question directly, they have often resorted to parading out a series of charts revealing only the total number of accidents due to aircrew error in very general terms. This is not to imply that the leadership of the U.S. Navy is evasive or that the analysts are incompetent. Rather, the numbers simply did not exist in a form that allowed such questions to be answered directly – at least, that is, until HFACS.

Shortly after the accident described above, the U.S. Navy began questioning the extent to which aviators in the fleet were involved in aviation accidents due to human error, particularly violations of the rules. Coincidentally, around that same time we had begun exploring the use of HFACS as a data analysis tool within the U.S. Navy/Marine Corps. As part of that effort, we systematically examined all of the human causal factors associated with Naval (both the U.S. Navy/Marine Corps) tactical aircraft (TACAIR)[8] and helicopter accidents that had occurred since 1991 (Shappell, et al., 1999). Little did we know that the results of our analyses would be the

---

[8] TACAIR includes U.S. fighter and attack aircraft like the A-4, A-6, AV/8B, C-2A, E-2, EA/6B, F/A18, F-5, F-14, KA/6D, and S-3.

impetus behind fundamental changes that would soon take place within Naval aviation.

During our investigation, we examined a variety of aircrew, supervisory, and organizational factors contained within 151 U.S. Naval Class A[9] aviation accident reports using a panel of experts including aerospace psychologists, flight surgeons, and Naval aviators. The experts were instructed to classify within HFACS, only those causal factors identified by the investigative board. In other words, they were not to "reinvestigate" the accident or second-guess the investigators and chain of command. To classify anything other than the official causal factors of the accidents would not only be presumptuous, but would only infuse opinion, conjecture, and guesswork into the analysis process and in so doing, threaten the credibility of the findings.

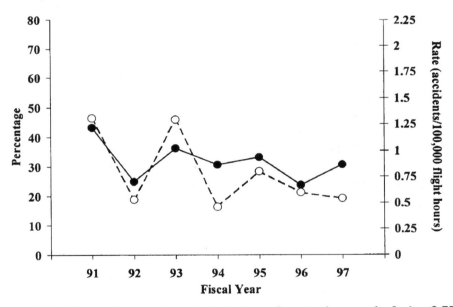

Figure 5.1 Percentage (closed circles) and rate (open circles) of U.S. Navy/Marine Corps Class A accidents associated with at least one violation as defined within HFACS

To many, the results of our analyses were alarming because we had discovered that roughly 1/3 of the Naval aviation accidents we examined

---

[9] The U.S. Navy/Marine Corps considers an accident as Class A if the total cost of property damage (including all aircraft damage) is $1,000,000 or greater; or a naval aircraft is destroyed or missing; or any fatality or permanent total disability occurs with direct involvement of Naval aircraft.

were associated with at least one violation of the rules (Figure 5.1). Regardless of whether one looked at the percentage of accidents associated with violations, or the rate (number of accidents associated with at least one violation per 100,000 flight hours) at which they occurred, the findings were the same. To make matters worse, the percentage and rate had remained relatively stable across the seven years of data we examined.

For obvious reasons, these findings did not sit well with the U.S. Navy/Marine Corps, and predictably, some within Naval leadership questioned the validity of our findings. After all, how could things be so bad? As a result, they sent their own experts to re-examine the data and much to their surprise, they got the same answer we did. There was no denying it now, the U.S. Navy/Marine Corps had a problem.

Faced with explaining how such a large percentage of accidents could be attributable, at least in part, to the willful disregard for the rules, some simply argued that this was the natural by-product of selecting people for military duty. That is, we intentionally recruit pilots who are willing to push the envelope of human capabilities to a point where they may ultimately be asked to lay down their lives for their country. Put simply, military aviation is filled with risks, and those who aspire to be Naval aviators are naturally risk-takers who, on occasion, may "bend" or even break the rules. While certainly a concern within the U.S. Navy/Marine Corps, those who were willing to accept this explanation argued that the U.S. Army and Air Force had the same problem.

Unfortunately, views like this were difficult, if not impossible to contest because the different branches of the military were using distinctly different investigative and archival systems, rather than a common human error framework. As a result, analysts were typically left with comparing little more than overall accident rates for each branch of the Armed Services rather than specific types of human errors. But with the development of HFACS, the U.S. military had a framework that would allow us to do just that!

We therefore set out to examine U.S. Army and Air Force Class A accidents using the HFACS framework (Wiegmann et al., 2002). The results of these analyses were very surprising not only to us, but to others in the Fleet as well, because we found that the percentage of accidents attributable to violations of the rules was *not* the same among the Services (Figure 5.2). Indeed, during roughly the same time frame, a little over one quarter of the Army and less than 10 percent of the Air Force accidents were associated with violations. Furthermore, these differences had nothing to do with the fact that unlike the U.S. Navy/Marine Corps, the Army almost exclusively flies helicopters and the Air Force tends to fly more point-to-point cargo and resupply missions. Keenly aware of these potential confounds, we intentionally compared "apples-to-apples" and "oranges-to-oranges." That is, we compared data from Army helicopter accidents with Naval helicopter accidents, and those involving Air Force TACAIR aircraft with their counterparts in the U.S. Navy/Marine Corps.

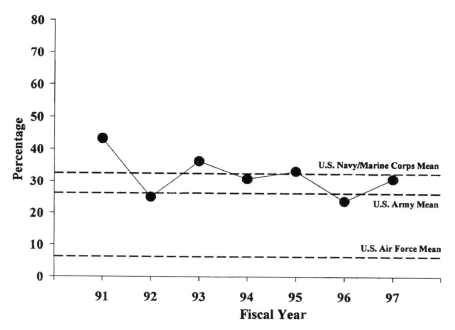

Figure 5.2 The percentage of U.S. Navy/Marine Corps Class A accidents associated with at least one violation as defined within HFACS. The mean percentages of Class A accidents for the U.S. Navy/Marine Corps, U.S. Army, and U.S. Air Force are plotted with dashed lines

Having run out of explanations and still faced with this previously unknown threat to aviation safety, senior leadership within the U.S. Navy/Marine Corps knew they had to do something – but what? How could Naval aviators be so much different than their counterparts in the other services? It turns out that the answer was right under our nose, or should we say right before our eyes on the television. For over a generation, Gene Roddenberry, the creator of *Star Trek*, entertained millions with the adventures of Captain Kirk and the crew of the starship *Enterprise*. Why that is relevant to our discussion is that he chose to model his futuristic Starfleet after the U.S. Navy for one simple reason. For nearly two centuries, the U.S. Navy had empowered their front line combatants (whether they were captains of seafaring vessels or pilots of aircraft) with the ability to make tactical decisions on their own. After all, when the U.S. Navy was first created, seafaring Captains could not simply "phone home" for permission to engage the enemy. Likewise, Gene Roddenberry, when creating his popular television series, wanted Captain Kirk to be able to engage the Klingons and other enemies of the Federation without having to call home for permission every time.

The Navy referred to this as "flexibility" and in many ways encouraged their officers to do what was necessary to get the job done and "beg forgiveness in the morning" if problems arose. It could even be argued that this "can-do" culture led to an attitude that if the rules did not explicitly say something could not be done it meant that you could do it. In stark contrast, within the U.S. Army and Air Force, if the rules did not say you could do something, it meant that you could *not* do it. As a result, if an aviator was caught "bending the rules" in the U.S. Navy/Marine Corps he would likely be taken behind the proverbial wood shed and spanked (figuratively, of course), but would then be allowed to return to tell everyone his story and fly another day. In the U.S. Army and Air Force, however, if you broke the rules you were immediately removed from duty.

When all was said and done, it looked to even the most skeptical observer that the U.S. Navy/Marine Corps had a problem whose roots were imbedded deep within the culture and traditions of the organization – at least where Naval aviation was concerned. Armed with this information, the U.S. Navy/Marine Corps developed and implemented a three-prong, data-driven intervention strategy that included a sense of professionalism, increased accountability, and enforcement of the rules.

To some, professionalism might sound a bit corny, but when you tell Naval Aviators that they are worse than their Army or Air Force counterparts, you quickly get their attention. Pilots, particularly those within the U.S. Navy/Marine Corps, are a very proud group. Any suggestion that they are anything less than professional, strikes at their very core, and in many ways provided the foundation for the other interventions that followed.

Now that they had the full attention of the Fleet, senior leadership began building a sense of accountability among their aircrews beyond what had been in place before. In other words, if a pilot willfully broke the rules, regardless of rank, there would be consequences, and in some instances removal from flight status. It did not take long before a couple of fighter pilots were caught violating established regulations and as a result, they were summarily removed from flight status. News, in the fleet travels quickly and soon everyone knew that senior leadership was serious. Not only that, but if it was discovered that the Commanding Officer or other squadron management were not enforcing the rules, they too would be held accountable. And soon, a couple of senior officers were relieved of duty as well.

Although some saw this new enforcement of the rules as Draconian, it was now clear to everyone that senior leadership was serious about reducing the number of accidents due to violations of the rules and regulations. The question was whether the interventions had any positive effect on the accident rate. Here again is where HFACS can help. Not only can HFACS identify human error trends in accident and incident data, but it is particularly useful for *tracking* the effectiveness of selective interventions as well.

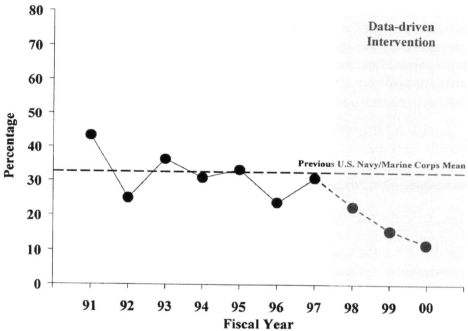

Figure 5.3 The percentage of U.S. Navy/Marine Corps Class A accidents associated with at least one violation in the years before and after (shaded region) the intervention strategy was implemented

Before the development of HFACS, the only way an organization could determine the effectiveness of a given intervention was to examine the accident rate without regard for specific types of human error. Unfortunately, overall accident rates are dependent on many things of which the targeted behavior is but one piece. Consequently, if the overall accident rate did not decline, one might be tempted to abandon the intervention, without knowing whether it actually worked. After all, if accidents due to violations decline, but some other factor is on the increase, the overall accident rate may not change appreciably. But with HFACS, we can monitor selective types of human error, not just an overall accident rate. So, with that in mind, Figure

5.3 is submitted as evidence that the U.S. Navy/Marine Corps has indeed made tremendous gains. That is, by 1998, the percentage of accidents associated with violations had been reduced to Army levels, and by 2000, they had nearly reached those seen in the U.S. Air Force. Proof positive that data-driven interventions implemented by the U.S. Navy/Marine Corps worked!

So, by using HFACS, the U.S. Navy/Marine Corps identified a threat to aviation safety, developed interventions, tracked their effectiveness, and "solved the problem" – or did they? As any psychologist or parent will tell you, the rules will be adhered to only as long as there is memory for the consequences. That being said, safety personnel within Naval aviation are keenly aware that as the current generation of pilots moves on, a new generation will take their place and we may once again see accidents due to violations increase. The good news is that we can track violations using HFACS, and if accidents due to this particular unsafe act begin to creep back up, the intervention can be re-evaluated and perhaps modified or reinforced.

Quantifying Proficiency within the Fleet

Throughout the last century, mankind has witnessed extraordinary advances within the world of aviation as propeller driven biplanes made of cloth and wood have been replaced by today's advanced turboprops and jets. Yet, improvements in aircraft design tell only part of the story as aviators have been forced to adapt with each innovation introduced to the cockpit. Gone are the celebrated days of the barnstorming pilots who flew as much by feel and instinct as they did by skill. Those pioneering aviators have been replaced by a generation of highly-educated technicians raised on Nintendo and computer games, flying state-of-the-art systems the likes of which the Wright Brothers could only dream of.

Yet, at what price has this apparent evolution among aviators taken place? Within the military, some have argued that while aircrew today are perhaps smarter and may even be better decision-makers, their basic flight skills cannot compare with previous generations of pilots. This is not to say that today's military aviators are poor pilots. On the contrary, they are extremely bright and uniquely talented. Still, when comparing them to their predecessors of even 20 years ago, today's military aircrews seem to rely as much on automation and advanced flight systems as they do on their own stick-and-rudder skills.

Are pilots today really that much different from those of generations past, or are such views simply the jaded opinions of those longing for the "good-old days" of aviation? Perhaps one way to answer this question is to examine the accident record for evidence that basic flight skills (otherwise known as proficiency) have eroded among today's aviators. That is, to the extent that accidents accurately portray the current state of the Fleet (and some may want to argue that point), one should be able to inspect the accident record for any evidence that would defend assertions that proficiency has degraded among military aircrews. More to the point, if proficiency has declined, the percentage of accidents associated with skill-based errors should naturally increase.

With this in mind, we examined the same U.S. Naval aviation accidents described above using the HFACS framework (Shappell and Wiegmann, 2000b). What we found is that over half (110 of 199 accidents, or 55 percent) were associated with skill-based errors. It is important to remember that we are *not* talking about complex decisions or misperceptions that may be easier to justify. No, these were simply breakdowns in so-called "monkey skills", those stick-and-rudder skills that we all take for granted. Even more disturbing is that the percentage of accidents associated with these errors has increased steadily since 1991 when just under half (43 percent) of the accidents were associated with skill-based errors. Yet, by the year 2000, an alarming 80% of the accidents were, at least in part, attributable to a breakdown in basic flight skills (Figure 5.4).

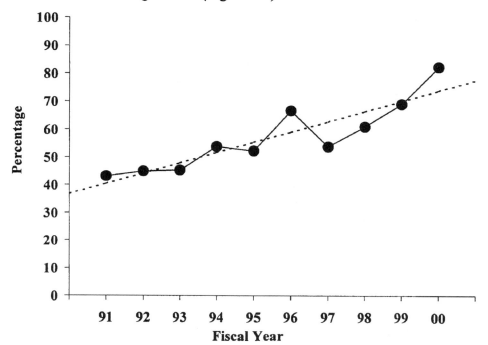

Figure 5.4 Percentage of accidents associated with skill-based errors. The linear trend is plotted as a dashed line

Given findings such as these, it would appear that there is at least some truth to the claim that aircrew proficiency has eroded over the decade of the 1990s – something that was little more than speculation before HFACS. So how could this happen to what many feel is the premiere military in the world? When searching for the answer, perhaps the best place to begin is with the pilots themselves. Indeed, if you ask most pilots, they will tell you that the erosion in basic flight skills was the direct result of a systematic reduction in flight hours after the Cold War. After all, flying, as with any skill, will begin to deteriorate if it is not practiced. Indeed, even the best athletes in the world cannot stay at the top of their game if they do not play regularly. It makes sense then that as flight hours decline, so to would basic flight skills. So, as any pilot will tell you, the solution is simple, provide more flight hours and proficiency will naturally improve.

Unfortunately, the answer may not be that straightforward. Some experts have suggested that the reduction in aircrew proficiency is directly related to the complexity of today's modern aircraft, making the issue of flight hours all the more important to the process of maintaining basic flight skills. For example, it has been said that it takes one and a half pilots to fly an F/A-18 Hornet, the U.S. Navy/Marine Corps' front line fighter aircraft. Recognizing this, the Marine Corps has chosen to fly the two-seat model and added a weapons systems operator to help manage the workload. The problem is, the Navy chose to use the single-seat version meaning that something had to give – but what?

An examination of the accident data revealed that together, single-seat and dual-seat Hornets account for roughly one quarter of all the aircrew-related accidents in the U.S. Navy/Marine Corps. Of these, just over 60 percent of the dual-seat Hornets have been associated with skill-based errors, similar to the percentage we have seen with other fleet aircraft. But what is particularly telling is that nearly 80 percent of the accidents involving single-seat Hornets have been attributed to skill-based errors. This finding lends some support to the belief that the complexity of modern fighter aircraft is a driving force behind the erosion of proficiency observed within Naval aviation. At a minimum, these data suggest that the single-seat version of the F/A-18 Hornet is likely to be the most sensitive to a reduction in flight hours.

While a reduction in flight time and the increasing complexity of military aircraft make for compelling arguments, others have proposed that over-reliance on automation may also be responsible for the erosion of basic flight skills seen among today's modern fighter pilots. Consider first that modern TACAIR aircraft are all equipped with autopilots that will maintain a variety of flight parameters (e.g., altitude, heading, and airspeed) at the push of a button. Now consider that a large part of the flight regime is flown using these sophisticated avionics, thus providing little opportunity for honing one's flight skills, particularly as new automated systems replace more and more phases of flight. It should come as no surprise then that basic flight skills learned very early in training would erode.

In contrast to modern fighter aircraft, most military helicopters still rely on manual flight controls where automation only serves to dampen or smooth out the inputs rather than maintain altitude, heading, or airspeed. Logically then, if one were to compare TACAIR pilots to helicopter pilots, marked differences among their basic flight skills should emerge. Not only that, but these differences should be consistent across the different branches of the Armed Forces.

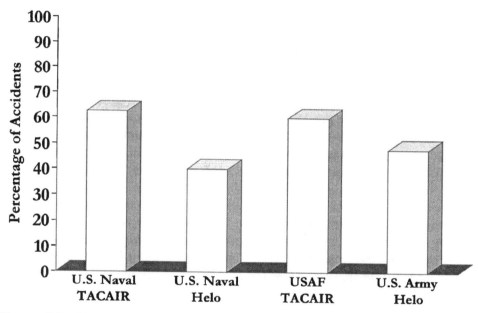

Figure 5.5 Percentage of U.S. military TACAIR and helicopter accidents occurring between FY 1991 and 2000 that were associated with skill-based errors

Again, the accident data may provide some support for this view as well. Because we now have a common framework for examining human error within U.S. military aviation, we can directly compare the percentage of skill-based errors associated with TACAIR and helicopter accidents over the last several years. As can be seen in Figure 5.5, significantly more TACAIR than helicopter accidents are associated with skill-based errors. Furthermore, this trend was consistent across the different Services as the percentage of accidents associated with skill-based errors was the same for Naval and Air Force TACAIR, as were those associated with Naval and Army helicopters.

Obviously, differences other that just automation exist between helicopters and TACAIR aircraft; but, regardless of the reasons, there is no denying that a smaller percentage of helicopter accidents are attributable to skill-based errors.

While it is easy to see how the reduction in flight hours, combined with the increased complexity of aircraft and even automation, may have led to the erosion in proficiency among Naval aviators, there may still be other explanations. Consider this opinion offered by a senior Naval aviator when comparing pilots today with those of his generation some 20-30 years earlier.

Pilots today are better educated than we were, make better tactical decisions, and fly the most advanced aircraft known to man. But while we may not have been as smart, or have flown sexy aircraft, we could fly circles around these guys.

There may be some truth to this point of view when you consider the emphasis placed on tactical decision-making during training. Much of the curriculum throughout the 1990s emphasized pilot decision-making, perhaps to the detriment of basic flight skills. Even today's modern flight simulators emphasize tactical simulations over the fundamental ability to fly the aircraft. The question is whether there is any support for this view in the accident data.

Indeed, there appears to be some evidence that the emphasis on decision-making has paid some dividends, at least where accidents are concerned (Figure 5.6). The percentage of accidents associated with decision errors has declined since 1991, and even more if we only consider 1994 to 2000. Recall that it was during these same years that we saw the increase in skill-based errors.[10] So, it does appear that while there has been a modest decline in the percentage of accidents associated with decision errors, it may have been at the expense of basic flight skills.

So, where does this leave us with the issue we began this section with, the erosion of proficiency among aircrew, in particular U.S. Naval aircrews? To the extent that accident data accurately reflects the state of Naval Aviation, it would appear that indeed proficiency has begun to erode. Although the debate continues over the exact causes of this troubling trend, everyone agrees that greater emphasis needs to be placed on simply flying the aircraft.

In response, the U.S. Navy/Marine Corps has embarked on a concerted effort to improve proficiency among its' aviators. For instance, they have

---

[10] Just because the percentage of one error form goes up from year-to-year does not necessarily mean something else must come down. Because there are multiple causal factors associated with most accidents, the percentages of each type of causal factor will not equal 100 percent. In this way, the various error forms are independent of one another.

recently instituted a "back-to-the-basics" approach that focuses on such issues as re-emphasizing the need for an efficient instrument scan, prioritizing attention, and refining basic flight skills. In addition, there are efforts underway to develop low-cost simulators that focus on basic stick-and-rudder skills and issues of proficiency. While such PC-based aviation training devices have been shown to be useful within the civilian sector, their use in military aviation is only now being explored. In the end, only time and the accident record will tell whether any of these interventions will prove successful in reversing this threat to Naval Aviation.

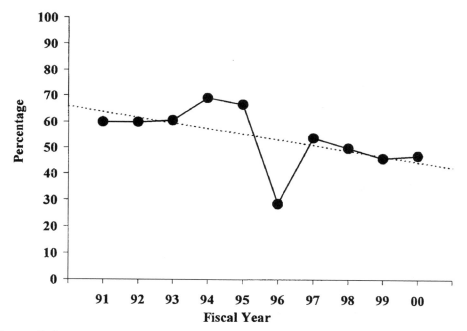

Figure 5.6 Percentage of accidents associated with decision errors. The linear trend is plotted as a dashed line

Crew Resource Management Training: Success or Failure?

With the pioneering work of Clay Foushee, Bob Helmreich, and Eduardo Salas, the role of aircrew coordination in aviation safety has taken center stage within the airline industry. Not surprising then, when Naval leadership discovered that several of their accidents were attributable to breakdowns in crew resource management (CRM), an approach for integrating CRM training into the Fleet was created. Introduced to a limited number of squadrons in the late 1980s, aircrew coordination training (ACT), as it came to be known, was based largely on leadership and assertiveness training developed by the airlines to address their own concerns with CRM. By the early 1990s, ACT had become fully integrated into both initial and recurrent Naval training and was expanded to cover such things as workload management and communication skills.

But was the U.S. Navy/Marine Corps' new ACT program effective at reducing the spate of accidents associated with CRM failures? Initial assessments of the program in 1992 were encouraging, particularly within those communities in which ACT was first introduced (Yacavone, 1993). As a result, it was widely believed that the program within the U.S. Navy/Marine Corps had been a huge success. So much so, that when faced with budget cuts after the Cold War, some officials suggested that funds allocated for ACT be redirected to other priorities like buying jet fuel or aircraft parts. After all, everyone had been trained on the principles of CRM and to paraphrase one senior Naval Officer, "we've got no [stinkin'] CRM problem in the Navy, at least not anymore." Unfortunately, follow-up analyses had not been conducted to assess the long-term impact of ACT on Naval aviation safety. This was due in large part to the difficulty of tracking CRM failures within the accident data and the growing belief that whatever CRM problems existed before, had been solved with ACT.

With the implementation of HFACS in the late 1990s, the Navy and Marine Corps were able to quickly and objectively assess the success of the ACT program. Regrettably, the results did not support early optimism. As illustrated in Figure 5.7, roughly 60 percent of Naval aviation accidents were found to be associated with a breakdown in CRM. Even more troubling, this percentage was virtually identical to the proportion seen in 1990, one year *before* the fleet-wide implementation of ACT. Arguably, there was a slight dip in 1992, which may explain the early enthusiasm regarding the flagship program. Since then however, the percentage of CRM accidents has fluctuated up and down, but ultimately has not declined, as many believed was the case.

Clearly, to even the most ardent supporter, the U.S. Navy's ACT program had not done what its original developers had hoped. Nevertheless, it did not make sense to shut down a program targeted at what appeared to be a large (60 percent of the accidents) and persistent (over 10 years) threat to aviation safety. On the other hand, to simply pour more money into an effort that yielded little improvement did not make much sense either. Instead, what the U.S. Navy/Marine Corps needed was a thorough evaluation of the ACT program so informed decisions could be made regarding the future of CRM training.

But, where does one start? After all, how could ACT have been so ineffective given the touted success of similar programs within commercial aviation? Therein lies at least part of the problem. After all, have commercial

CRM programs really been successful? Or, are such claims based simply on anecdotal and other subjective evidence presented by those too close to the issue to remain objective?

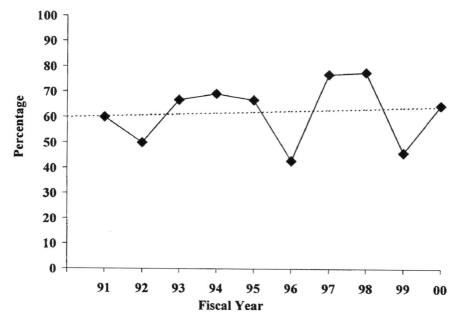

Figure 5.7 Percentage of accidents associated with crew resource management failures. The linear trend is plotted as a dashed line

With the transition of the HFACS framework from the military to the civilian sector, we were in a unique position to directly address this question. Using scheduled air carrier[11] accidents occurring between 1991 and 1997, we found that roughly 30 percent of all air carrier accidents were associated with a breakdown in CRM, a proportion much lower than that seen in the U.S. Navy/Marine Corps (Figure 5.8).

At first glance, this might lead some to conclude that the commercial program had been a success, at least when compared with Naval aviation. However, a closer inspection of the graph revealed that like the Navy and Marine Corps, the percentage of accidents associated with CRM failures has

---

[11] The data in this analysis used 14 Code of Federal Regulations (CFR) Part 121 and Part 135 scheduled air carrier accidents. The designation 14 CFR Part 121 pertains to those domestic, flag, and supplemental operations holding an Air Carrier Certificate, having 30 or more seats, and a maximum payload of more than 7,500 pounds. The designation 14 CFR Part 135 refers to commuter or on-demand operations.

remained largely unchanged since 1991. Worse yet, the percentage may have even increased slightly.

Notably, we were not the only ones that that identified this disturbing trend within the accident data. The Government Accounting Office (GAO) had initiated a major review of the commercial CRM program and concluded essentially the same thing (GAO, 1997). That is, commercial CRM programs have been largely ineffective at reducing accidents due to CRM failures. Thus, it would appear that our analyses had been validated (but, one thing is certain; given our current salaries, the HFACS analyses were much cheaper to perform).

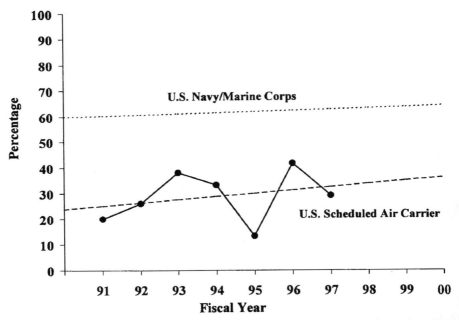

Figure 5.8 Percentage of U.S. scheduled air carrier accidents associated with crew resource management failures. The linear trends for the U.S. Navy/Marine Corps and scheduled air carrier accidents are plotted as dashed lines

Now, before our colleagues start sending us hate mail for criticizing their CRM programs, we admit that CRM has changed considerably since it was first introduced to commercial aviation nearly two decades ago. In fact, those who pioneered the field were the first to recognize the limitations of the initial strategy for teaching CRM (for a brief review see Salas et al., 2003). For instance, many of the early programs amounted to little more than personality assessment and classroom discussions of high profile case studies with a little bit of science thrown in for good measure.

Yet, even the best classroom instruction is limited if participants are not required to demonstrate the principles they have learned. In other words, simply making people aware that they need to communicate better without compelling them to practice in a real-world situation (e.g., in the aircraft or simulators) is tantamount to telling student pilots how to fly an airplane and expecting them to be able to do it the first time without practice!

To address this limitation, programs like line-oriented flight training (LOFT) were developed in the early 1990s, requiring pilots to demonstrate, in a flight simulator, the skills they learned in the classroom. More recently, programs like the Advanced Qualification Program (AQP), have integrated CRM training and evaluations into the actual cockpit. A voluntary program employed by nearly all major air carriers in the U.S. and a large portion of regional airlines, AQP requires that both initial and recurrent qualifications include inflight evaluations of specific technical and CRM skills. Although these new training programs are promising, whether they improve CRM and reduce accidents remains to be seen.

Recognizing the potential benefit of new programs like LOFT, the U.S. Navy began experimenting with their own version in 1997. In particular, a few S-3 Viking squadrons began using videotape feedback and simulators as part of their recurrent training program. Preliminary results (albeit subjective in nature) have been positive; but whether the program will be expanded to include the entire Fleet has yet to be determined.

Perhaps more important in the near term, was the realization that the ACT curriculum had not been tailored to meet the specific needs of the targeted community. Indeed, much of the curriculum had yet to evolve beyond a few "classic" examples of civilian aviation accidents involving CRM failures. In essence, instructors were teaching combat pilots lessons learned from accidents like the Eastern Air Lines L-1011 that crashed into the Florida Everglades (NTSB, 1973). While the movies may have been informative, it is hard for an F/A-18 Hornet pilot to make the connection between a burned out light bulb on a commercial airliner and CRM failures in one of the most advanced fighter jets in the world today. On top of everything else, most of the examples were outdated, narrow in scope, and did not capture the factors that contribute to CRM failures in Naval aviation.

After realizing the folly of their ways, it was no surprise to Naval leadership that ACT has had little or no impact in the Fleet. As a result, a concerted effort was undertaken to identify platform-specific CRM accidents as teaching tools for future CRM training. That is, the U.S. Navy/Marine Corps redesigned the program to teach F/A-18 CRM to F/A-18 pilots and F-14 CRM to F-14 pilots. Using specific examples relevant to each community, the goal was to identify those aspects that are unique to single-seat versus multi-seat aircraft, fighter versus attack, fixed-wing versus helicopter, and any other nuances that are relevant to today's modern Naval aviator. With HFACS, Naval leadership will be able to accurately and objectively assess those aspects that work to reduce CRM related accidents and those that do not.

The Redheaded Stepchild of Aviation

Understandably, a great deal of effort has been expended over the last several decades to improve safety in both military and commercial aviation. Yet, even though hundreds of people have died and millions of dollars in assets have been lost in these operations, the numbers pale by comparison to those suffered every year within general aviation (GA). Consider the decade of the 90's. For every commercial or military accident that occurred in the U.S., roughly nine GA aircraft crashed (Table 5.1). More alarming, nearly one in five GA accidents (roughly 400 per year) involved fatalities – resulting in a staggering 7,074 deaths! Since 1990, no other form of aviation has taken more lives.

Table 5.1 The number of accidents annually for U.S. commercial, military, and general aviation

Year	Commercial	Navy/ Marine Corps	Army	Air Force	Totals	General Aviation
1990	146	63	32	51	292	2,241
1991	137	59	43	41	280	2,197
1992	117	57	25	48	247	2,111
1993	108	43	22	34	207	2,063
1994	118	31	15	35	199	2,022
1995	125	30	10	32	197	2,056
1996	138	40	8	27	213	1,908
1997	147	25	16	29	217	1,845
1998	135	33	10	24	202	1,904
1999	138	27	18	30	213	1,906
Totals	1,309	408	199	351	2,267	20,253

Source: *U.S. Naval Safety Center, U.S. Army Safety Center, U.S. Air Force Safety Center, NTSB.*

Why then has general aviation received so little attention? Perhaps it has something to do with the fact that flying has become relatively commonplace, as literally millions of travelers board aircraft daily to get from point A to point B. Not surprising then, when a commercial airliner does go down, it instantly becomes headline news, shaking the confidence of the flying public. To avoid this, the government has focused the bulk of their limited aviation resources on improving commercial aviation safety.

But does the commercial accident record warrant the lion's share of the attention it has received? Well, if you consider the data in Table 5.1, there are about 130 commercial aircraft accidents per year. However, of these 130 so-called "accidents," many are simply injuries in the cabin due to turbulence or involve small, on-demand air taxis. Thankfully, very few are on the scale of TWA Flight 800, the Boeing 747 that crashed off the coast of New York in July of 1996 killing all 230 passengers. In fact, of the 1,309 commercial airline accidents that occurred in the 1990s, only a handful involved major air carriers and fewer yet were associated with fatalities.

On the other hand, in the time it took you to read this book there have probably been 20 GA accidents in the U.S. alone, of which four involved deaths. But did you hear about any of them on the primetime evening news, or read about them on the front page of *USA Today*. Probably not, after all they happen in isolated places, involving only a couple of hapless souls at a time. In fact, unless the plane crashed into a school, church, or some other public venue, or involved a famous person, it is very unlikely that anyone outside the government or those intimately involved with the accident even knew it happened.

Although GA safety may not be on the cusp of public consciousness, a number of studies of GA accidents have been conducted in an attempt to understand their causes. Unfortunately, most of these efforts have focused on contextual factors or pilot demographics rather than the underlying cause of the accident. While no one disagrees that contextual factors like weather (e.g., IMC versus VMC), lighting (e.g., day versus night), and terrain (e.g., mountainous versus featureless) contribute to accidents, pilots have little, if any, control over them. Likewise, knowing a pilot's gender, age, occupation, or flight experience, contributes little to our ability to prevent GA accidents. After all, just because males may have a higher accident rate than females, are we now going to prohibit men from flying? Or how about this well publicized bit of trivia: pilots with fewer that 500 flight hours have a higher risk of accidents. What are we as safety professionals to do, wave a magic wand in the air and *poof*, a 300-hour pilot can now fly with the prowess of someone with 1000 flight hours under their belt? Truth be told, this information has provided little in the way of preventing accidents apart from identifying target audiences for the dissemination of safety information.

In fact, even when human error has been addressed, it is often simply to report the percentage of accidents associated with aircrew error in general or to identify those where alcohol or drug use occurred. Recently however, we examined over 14,500 GA accidents using five independent raters (all were certified flight instructors with over 3,500 flight hours) and the HFACS framework. What we found was quite revealing, as previously unknown error trends among general aviation were identified.

Let us first look at the roughly 3,200 fatal GA accidents associated with aircrew error. From the graph in Figure 5.9, some important observations can be made. For instance, it may surprise some that skill-based errors, not decision errors, were the number one type of human error associated with fatal GA accidents. In fact, accidents associated with skill-based errors (averaging roughly 82 percent across the years of the study) more than doubled those seen with decision errors (36 percent) and violations of the rules (32 percent). Even perceptual errors, the focus of a great deal of interest over the years, were associated with less than 12 percent of all fatal accidents.

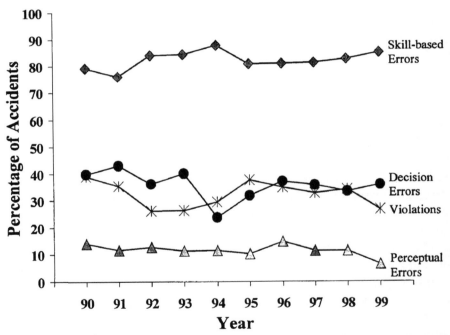

Figure 5.9 Percentage of fatal GA accidents associated with each unsafe act (skill-based errors – diamonds; violations – asterisks; decision errors – filled circles; perceptual errors – triangles)

Also noteworthy is the observation that the trend lines are essentially flat. This would seem to suggest that safety efforts directed at general aviation over the last several years have had little impact on any specific type of human error. If anything, there may have been a general, across-the-board effect, although this seems unlikely given the safety initiatives employed.

The only exceptions seemed to be a small dip in the percentage of accidents associated with decision errors in 1994 and 1995 and a gradual decline in those associated with violations between 1991–94. In both cases however, the trends quickly re-established themselves at levels consistent with the overall average.

While this is certainly important information, some may wonder how these findings compare with the nearly 11,000 non-fatal accidents. As can be seen in Figure 5.10, the results were strikingly similar to those associated with fatalities. Again, the trends across the years were relatively flat and as with fatal accidents, skill-based errors were associated with more non-fatal accidents than any other error form, followed by decision errors, violations, and perceptual errors respectively.

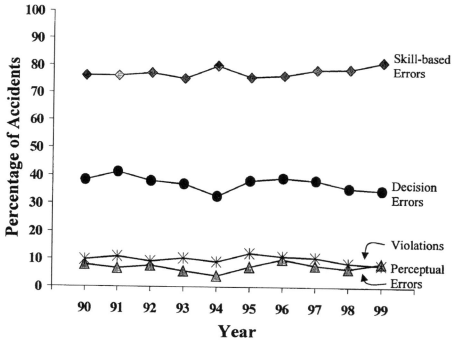

Figure 5.10 Percentage of nonfatal GA accidents associated with each unsafe act (skill-based errors – diamonds; violations – asterisks; decision errors – filled circles; perceptual errors – triangles)

While the similarities are interesting, it was the differences, or should we say, the difference, that was arguably the most important finding. When the error trends are plotted together for fatal and non-fatal GA accidents, as they are in Figure 5.11, it is readily apparent that the proportion of accidents associated with violations was considerably less for non-fatal than fatal GA accidents. In fact, using a common estimate of risk known as the odds ratio, fatal accidents were more than four times more likely to be associated with violations than non-fatal accidents (odds ratio = 4.314; 95 percent confidence interval = 3.919 to 4.749, Mantzel-Haenszel test for homogeneity = 985.199, p<.001). Put simply, if a violation of the rules resulting in an accident occurs, you are considerably more likely to die or kill someone else than get up and walk away.

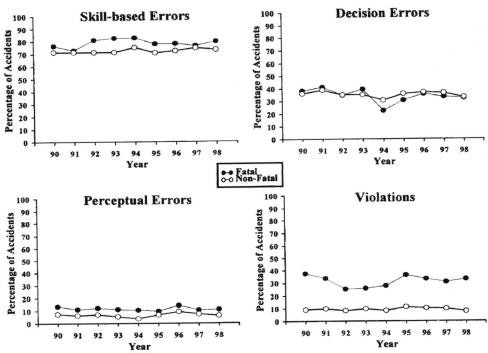

Figure 5.11 Percentage of fatal (closed-circles) and nonfatal (open circles) GA accidents associated with each unsafe act

So, what does all this mean? For the first time ever, we can talk about more than just the fact that nearly 80 percent of all general aviation accidents are attributable to "human error." After all, would you continue see a physician who only confirmed that you were "sick" without telling you what you what was wrong or what was needed to make you better? Probably not. The good news is that we now know "what is wrong" with general aviation – at least from a human error point of view. Specifically, the vast majority of GA accidents, regardless of severity, are due to skill-based errors. Also evident from our analyses, was the observation that one-third of all fatal accidents are due to violations of the rules and they are much less common in non-fatal accidents.

All of this leads to the inevitable question, "what can be done now that the face of human error has been exposed within general aviation?" Well, the data does suggest some possible avenues for preventing accidents. For example, there is a need to address the large percentage of accidents associated with skill-based errors. Perhaps placing an increased emphasis on refining basic flight skills during initial and recurrent flight training could possibly be effective in reducing skill-based errors. However, if the goal is to reduce fatal accidents, then greater emphasis must also be placed on reducing the number of violations through improved flight training, safety awareness, and enforcement of the rules. Still, before such interventions can be effectively applied, several other questions concerning the nature and role of human error in aviation accidents need to be addressed.

Conclusion

Using a human error framework like HFACS allows safety professionals and analysts alike to get beyond simply discussing accidents in terms of the percentage and rate of human error in general. Instead, we can now talk about specific types of human error, thereby increasing the likelihood that meaningful and successful intervention strategies can be developed, implemented, and tracked.

Imagine where we would be today if all we were still talking in generalities about mechanical failures without describing what part of the aircraft failed and why. Indeed, we might still be losing aircraft and aircrew at rates seen in the 1950s! Why then should we expect any less from human error investigations and analyses?

The question remains, with so many human error frameworks available, how do you know which one is best for your organization. After all, we don't know of any developers or academicians who believe that their error framework is of little use in the lab or in the field. They all feel good about their frameworks. But as consumers, safety professionals should not have to rely on the developers for confirmation of a particular error framework's worth. Ideally, we would have some criteria that can be used to help us decide. With that in mind, let us turn to the next chapter.

6 Beyond Gut Feelings...

Clearly, HFACS or any other framework adds little to an already long list of human error taxonomies if it does not prove useful in the operational setting. For that reason, we have made a concerted effort throughout the development process to ensure that it would have utility not only as a data analysis tool, but as a structure for accident investigation as well. The last thing we wanted was for HFACS to be merely an academic exercise applauded by those perched in their ivory towers, but of little use in the real world.

In a sense then, we were serving two masters when we began our work. On the one hand, as accident investigators ourselves, we wanted to ensure that our colleagues in the field could use HFACS to investigate the human factors associated with aviation incidents and accidents. On the other hand, as academic wolves in sheep's clothing, we wanted to make certain that it could withstand the scientific scrutiny that would inevitably come.

While we would like to think that we did a reasonable job straddling the fence as it were, how does one really know? Is there anything other than opinions upon which to base an evaluation of HFACS, or for that matter, any other framework? After all, error analysis systems like HFACS were designed to eliminate intuition and "gut feelings" from the accident investigation process. To turn around and evaluate their worth using those same gut feelings seems hypocritical at best and intellectually dishonest at worst. No, what we needed was to get beyond gut feelings.

The good news is that HFACS was developed using an explicit set of design criteria well known within the research community, but perhaps not as familiar elsewhere (Wiegmann and Shappell, 2001a). The purpose of this chapter therefore is to describe those criteria within the context of the development of HFACS' development to illustrate our points. Our hope is that safety professionals and others will consider these criteria when deciding which error framework to use within their own organizations.

Before we begin, however, we want to warn the reader that this chapter is by far the most academic (a code word for boring) in the book. So, if you are not the type that really enjoys reading scientific articles or talking statistics, then this is definitely a chapter for you to skim. On the other hand, if you are interested in learning more about the scientific rigor involved in developing and testing error analysis systems like HFACS, you may want to read a bit more closely. We will catch up with, or perhaps wake up, the rest of you in Chapter 7.

Validity of a Framework

The utility of any human error framework centers about its validity. That is to say, *what* exactly does the taxonomy measure, and *how well* does it do so (Anastasi, 1988). Yet, as important as validity is, surprisingly little work has been done to examine or compare the validity of different error models in an applied context. This may be due, in part, to the fact that assessing the validity of a framework can be a very difficult and overwhelming task for even the most scholarly of us. Indeed, several methodological approaches for testing the validity of analytical techniques and measurement tools have been proposed (Anastasi, 1988). But, while there are many types of validity (Figure 6.1), and even more ways to measure it, three types (content, face, and construct validity) are essential if an error taxonomy is going to be useful in the field.

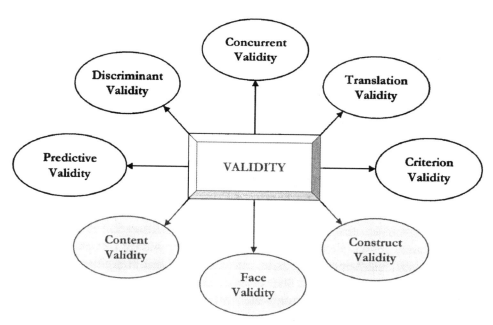

Figure 6.1 Types of validity with those relevant to error taxonomies highlighted

The first, *content validity* refers to whether a given framework adequately covers the error domain to be measured (Carmines and Zeller, 1979). That is, does the system capture the multitude of ways a human can err, or does it leave some important factors out. As we discussed in Chapter 2, many error taxonomies do not address the important precursors to human error. For instance, some focus entirely on the operator, ignoring the role played by supervisors and the organization in the genesis of human error. These so-called "blame-and-train" systems suffer from a lack of content validity in addition to a variety of other, perhaps more obvious, problems.

Face validity, on the other hand, refers to whether a taxonomy "looks valid" to those who would use it or decide whether or not to adopt its use within their organization (Carmines and Zeller, 1979). For example, does a particular framework seem like a reasonable approach for identifying the human factors associated with aviation accidents? More specifically, does it appear well designed, and will it work reliably when employed in the field by *all* levels of investigators? If the answer to any of these questions is "no," then the system is said to lack face validity.

Finally, *construct validity*, seeks to bridge the gap between a theoretical concept (e.g., human error) and a particular measuring device or procedure like HFACS (Carmines and Zeller, 1979). In other words, to what extent does a framework tap into the underlying causes of errors and accidents? Does the system really address why accidents occur, or does it simply describe what happened or restate the facts. Hopefully, this sounds familiar since it is the same argument that we made in earlier chapters against merely describing what occurred (e.g., the pilot failed to lower the landing gear), without identifying why (e.g., mental fatigue, distraction, etc.). In many ways then, construct validity, although difficult to evaluate, is the most important form of validity associated with error taxonomies.

Factors Affecting Validity

A number of theorists have proposed objective criteria for inferring the validity of error frameworks in applied settings (Hollnagel, 1998; O'Connor and Hardiman, 1996). However, these criteria vary widely across researchers and domains. That being said, a recent review of the literature (Wiegmann and Shappell, 2001a) suggests that at least four factors need to be considered when evaluating an error framework. As illustrated in Figure 6.2, these include: reliability, comprehensiveness, diagnosticity, and usability. We will address each of these in turn over the next several pages.

Reliability

If an error framework is going to be practical, users must be able to identify similar causal factors and reach the same conclusions during the course of an accident/incident investigation (O'Connor and Hardiman, 1996). Unfortunately, complete agreement among investigators has proven virtually impossible using existing accident investigation schemes. As a result, developers of human error frameworks are continually striving to improve the reliability of their systems.

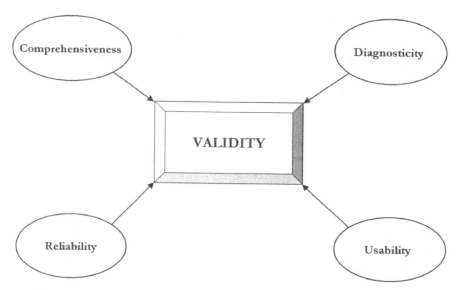

Figure 6.2 Factors affecting the validity of an error-classification system

Assessing reliability. As illustrated in Figure 6.3, the practice of assessing and improving the reliability of a classification system is an iterative process involving several steps. To begin, one must first decide on what level of agreement (i.e., inter-rater reliability) is acceptable. Ideally, that would be 100 percent agreement. However, as we just mentioned, reality dictates that something less will be required since very few, if any, classification systems will yield 100 percent agreement all of the time – particularly when human error is involved.

After deciding upon a suitable level of inter-rater reliability, the investigation/classification system can then be tested using independent raters and a sample data set to determine if the error framework produces the requisite level of inter-rater reliability. If not, modifications may need to be made to the error categories, definitions, instructions, or other areas to improve agreement among the raters.

After the necessary changes have been made, the new (presumably improved) taxonomy can then be applied to another sample of accident data, and inter-rater reliability reassessed. If the reliability has reached acceptable levels, the process stops. If not, the iterative process will continue until the target levels are eventually attained.

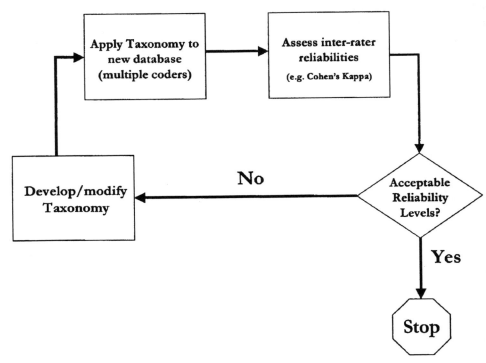

Figure 6.3 The process of testing and improving the reliability of an error classification system

Measuring reliability. While the process seems simple on the surface, let us assure you it is not. Even deciding upon an adequate measure of inter-rater reliability can be difficult since there are many ways to quantify it. Perhaps the most common method is to simply calculate the percentage of agreement among independent raters for a given classification task. For instance, if two raters agreed on 170 out of 200 classifications, they would have agreed 170/200 = 0.85 or 85 percent of the time.

The problem is that people will agree some of the time simply by chance. Consider, for example, a classification system with just two categories (e.g., errors of omission and errors of commission). For any given causal factor, there is a 25 percent chance that you would find agreement even if both raters simply guessed. Let us explain. The probability that Rater A classified a given causal factor as an "error of omission" by chance alone is 1 in 2 or 0.50. Likewise, the probability that Rater B classified the same causal factor as an "error of omission" by chance is 0.50, as well. To obtain the probability that both raters agreed simply by chance, you just multiply the two probabilities (0.50 × 0.50) and get 0.25. In other words, 25 percent of the time Raters A and B would be expected to agree by chance alone.

Beyond Gut Feelings... 127

To control for this, a more conservative statistical measure of inter-rater reliability, known as Cohen's *Kappa*, is typically used (Primavara et al., 1996). Cohen's *Kappa* measures the level of agreement between raters in excess of the agreement that would have been obtained simply by chance. The value of the *kappa* coefficient ranges from one, if there is perfect agreement, to zero, if all agreements occurred by chance alone. In general, a *Kappa* value of 0.60 to 0.74 is considered "good" with values in excess of 0.75 viewed as "excellent" levels of agreement (Fleiss, 1981). At a minimum then, the goal of any classification system should be 0.60 or better.

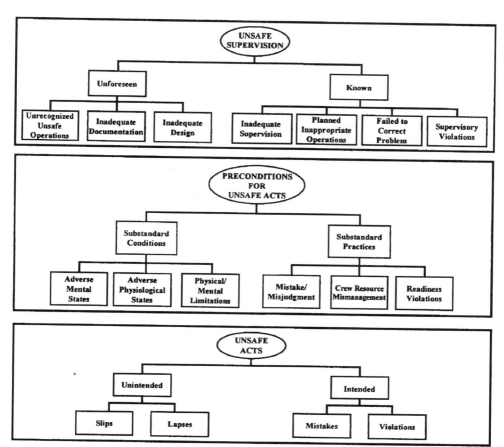

Figure 6.4 The Taxonomy of Unsafe Operations

To illustrate how testing and improving reliability can perfect an error framework, we will walk you through our efforts during the development of HFACS. You may recall that our first attempt at designing a human error framework resulted in the *Taxonomy of Unsafe Operations*[12] (Figure 6.4), a system based largely on the approach described by Reason (1990). While causal factors from actual Naval aviation accidents served as "seed data" throughout the development process, our intentions were largely academic, with little thought given to the taxonomy's potential use by accident investigators (Shappell and Wiegmann, 1995; Wiegmann and Shappell, 1995). Yet, by 1996, it became apparent that our work had possibilities beyond simply analyzing existing accident data. Indeed, the U.S. Navy/Marine Corps had expressed interest in using the taxonomy in the field. But, could it be reliably used by non-academics since most accident investigators in the field do not possess Ph.D.s in human factors?

To answer this question, Walker (1996) and Rabbe (1996) examined 93 U.S. Navy/Marine Corps controlled flight into terrain (CFIT) accidents and 79 U.S. Air Force F-16 accidents respectively, using the *Taxonomy of Unsafe Operations*. Employing pilots as raters, rather than academicians, the two studies classified over 700 causal factors associated with a combined 172 accidents. Their findings revealed that while the overall reliability among the raters was considered "good" using Cohen's *Kappa* (Table 6.1), inter-rater reliability was best for causal categories within the preconditions for unsafe acts, with slightly lower results for categories within the unsafe acts and unsafe supervision tiers.

Table 6.1 Reliability of the HFACS framework using military accident data

Author	Cohen's *Kappa*		
	Rater 1 vs Rater 2	Rater 1 vs Rater 3	Rater 2 vs Rater 3
Walker (1996)	0.70	0.60	0.65
Rabbe (1996)	0.69	0.78	0.62
Ranger (1997)	0.81	0.69	0.80
Plourde (1997)	0.89	0.85	0.86
Johnson (1997)	0.93	0.95	0.95

Based on these findings, a number of modifications were made to the taxonomy using input from accident investigators and the scientific literature (Figure 6.5). Initially, our focus was on the unsafe acts of operators. In many ways, we were forced to go back to the drawing board when it became clear that the concept of intended and unintended acts was lost on Walker and Rabbe's pilot-raters. For those that may not be familiar with Reason's (1990) description of intended and unintended actions, he felt that any description of unsafe acts must first consider the intentions of those

---

[12] For a complete description of the *Taxonomy of Unsafe Operations*, see Shappell and Wiegmann (1997a).

committing them. In other words, "Did the action proceed as planned?" If so, then the behavior is considered intentional; if not, it is considered unintentional. Unfortunately, trying to determine the intentions of aircrew involved in accidents (particularly those that perished in the crash) proved extremely difficult as evident from the modest levels of agreement between the pilot-raters.

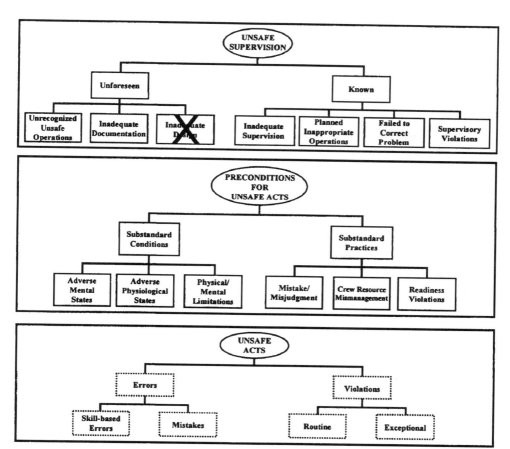

Figure 6.5 Modifications made to the *Taxonomy of Unsafe Operations*. Boxes outlined in dashes represent category changes. Categories deleted are indicated with an "X"

Therefore, after discussions with several pilots and aviation accident investigators, what seemed to make more sense was to distinguish between errors and violations as described in Chapter 3. We therefore restructured the unsafe acts around those two overarching categories.

But, the concept of intended and unintended actions was not the only problem that the pilot-raters had with the original *Taxonomy of Unsafe*

Operations. The distinction between slips and lapses was also difficult for the pilot-raters to understand and manage. Slips, as described by Reason (1990), are characteristic of attention failures and take the form of inadvertent activations, interference errors, omissions following interruptions, and order reversals, among others. In contrast, lapses typically arise from memory failures and include errors such as omitted items in a checklist, place losing, and forgotten intentions.

If all this sounds familiar, it should, since both slips and lapses comprise a large part of the skill-based errors category described in Chapter 3, and certainly made more sense to our investigators. We therefore abandoned the categories of slips and lapses and adopted skill-based errors as one of two types of errors committed by operators. The other type, mistakes, was kept in its entirety.

We also separated the category of violations into routine and exceptional violations as described in Chapter 3 and made some minor adjustments to the definitions of the remaining categories. Finally, we combined inadequate design with unrecognized unsafe operations since both were considered "unknown/unrecognized" by supervisors or those within middle management.

Two additional studies were then conducted to determine what, if any, impact the changes had on inter-rater reliabilities. Using the revised taxonomy, Ranger (1997) and Plourde (1997) examined causal factors associated with 132 U.S. Navy TACAIR and rotary wing accidents and 41 B-1, B52, F-111, and F-4 accidents, respectively. Findings from those studies revealed sizeable increases in agreement over those found by Walker (1996) and Rabbe (1996) the previous year (Table 6.1).

Nevertheless, while we were close, we still felt that there was room for improvement. We therefore made additional modifications to the framework as illustrated in Figure 6.6. Essentially, we added the category of perceptual errors and renamed mistakes to decision errors within the unsafe acts tier. In addition, we deleted the category of mistakes/misjudgment from the preconditions for unsafe acts because it was often confused and not utilized by the pilot-raters. Likewise, the unforeseen supervisory failures were deleted in large part because the Navy felt that if something was unforeseen, it was difficult to hold supervisors culpable since even Navy leaders cannot know everything.

Yet another reliability study was then conducted using what was renamed the Failure Analysis and Classification System (FACS)[13] and 77 U.S. Air

---

[13] To give credit where credit is due, FACS was later changed to the Human Factors Analysis and Classification System (HFACS) by COL Roger Daugherty, USMC and CAPT James Fraser who felt that FACS would be confused with the Marines' use of the acronym for forward air controllers.

Force A-10 accidents (Johnson, 1997). Overall, pair-wise reliabilities were found to be excellent (*Kappa* = 0.94) and consistent across all levels (Table 6.1). So, it would appear that by the end of 1997, we had developed a highly reliable error framework for use in the field as well as an analytical tool. But, would it be useful outside the military – that remained to be seen.

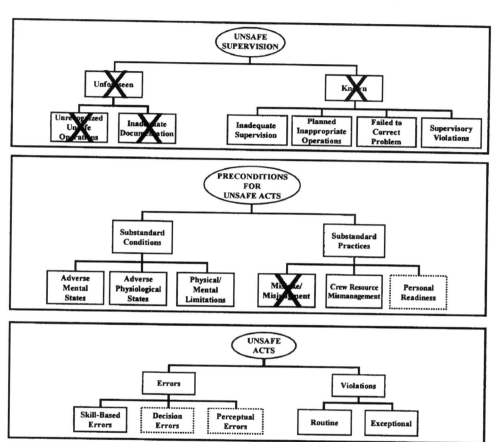

Figure 6.6 Additional modifications made to the *Taxonomy of Unsafe Operations*. Boxes outlined in dashes represent category changes. Categories deleted are indicated with an "X"

Early in 1998, we were interested in whether reliability would suffer if HFACS were used with civilian aviation accidents, given that our error framework was originally developed for use within the military. We were also curious whether the newly developed organizational influences tier would enjoy the same degree of inter-rater reliability as the rest of HFACS. We therefore conducted two additional reliability studies using commercial aviation accident data (Wiegmann et al., 2000). The first of these studies involved 44 scheduled air carrier accidents occurring between January 1990 and December 1997.

The overall reliability of the HFACS coding system was again assessed by calculating inter-rater reliability between our pilot-raters. Overall, the independent raters agreed 73 percent of the time yielding a *Kappa* value of 0.65. As expected, the highest level of inter-rater agreement was found among the unsafe acts and preconditions for unsafe acts, while the lowest level of agreement was within the supervisory level and the newly developed organizational tier.

Post-analysis discussions with raters suggested that the definitions and examples used to describe HFACS were too closely tied to military aviation and therefore somewhat ambiguous to those without a military background. For example, the concept of "flat-hatting," while commonly used within the U.S. Navy/Marine Corps to describe a Naval aviator who recklessly violates the rules to impress family and friends, was foreign to our civilian raters. To remedy this, the tables and checklists were modified to include more civilian examples, and yet another study was conducted to reassess the reliability of the HFACS system.

The follow-up study involved a new set of 79 commercial aviation accidents. This time, the two independent raters agreed roughly 80% of the time for a *Kappa* value of 0.75, considered "excellent" by conventional standards. Still, not satisfied by such a small sample size, we subjected the HFACS framework to the largest reliability test yet. Using over 2,500 general aviation accidents associated with more than 6,000 causal factors, five independent pilot-raters agreed more than 80 percent of the time yielding a *Kappa* value of 0.72 (Wiegmann and Shappell, 2001c). Given the fact that most safety professionals agree that the data associated with GA accidents is sparse and often cryptic, we were quite surprised and obviously pleased with the findings. Furthermore, these percentages varied only slightly across the years examined in this study (the actual range was 77 percent to 83 percent agreement).

To summarize then, over the course of nearly five years of research and development, HFACS evolved from little more than an academic exercise to a full-fledged accident investigation and analysis tool used by both military and civilian organizations. What's more, throughout the entire development process, we utilized the power of statistics and data analysis to ensure that inter-rater reliability was maximized knowing that a wide variety of individuals, with disparate educations and backgrounds would use HFACS.

Comprehensiveness

Reliability is only one of four criteria essential to the validity of any error framework. Comprehensiveness, or the extent to which a framework captures all the key information surrounding an accident, is important to the validity of any error classification system as well (O'Connor and Hardiman, 1996). However, this definition may be a bit misleading since a framework's comprehensiveness is really in the eye of the beholder. For instance, if one's interest is only in aircrew error, a cognitive framework may suffice. After all, when it comes to identifying and classifying information processing errors (e.g., decision errors, perceptual errors, attention errors, etc.), cognitive models tend to be very detailed and complete. Unfortunately, if your interest extends beyond operator error to include other contextual and organizational factors, these models tend to fall short, which may explain why they are rarely ever used exclusively.

Indeed, most organizations are interested in the breadth of possible error causal factors (e.g., supervisory, working conditions, policies, procedures, etc.), in addition to operator error. For that reason, organizational frameworks may be more suitable. Regrettably, as discussed in Chapter 2, these frameworks are not without their limitations either. Often, they sacrifice the detail seen with cognitive models making them more comprehensive globally, but less so on any one dimension.

In a perfect world though, one would not have to choose between breadth and depth in an error framework. A truly comprehensive system should be able to capture *all* the relevant variables that a given organization is interested in pursuing and perhaps even some that it may not. After all, it is hard to predict what aspects of human error an organization will actually be interested in tracking from year-to-year, much less 10 to 20 years from now. No one wants to find out that the framework they have invested years of effort and thousands of dollars in lacks the detail necessary to answer the safety questions at hand.

This brings us to yet another issue regarding error frameworks: *comprehensive* does not necessarily mean large. One has to distinguish between a "database" that is primarily used for archival purposes, and an error framework that is used for accident investigation and data analysis. Consider, for example, databases like ICAO's ADREP-2000. While it is certainly comprehensive, its overwhelming size makes it of little use for data analysis (Cacciabue, 2000). In fact, some have even quipped that the number 2000 in ADREP's title stands for the number of causal factors in the database, rather than the year it was created! In contrast, error frameworks like HFACS are not databases per se, but are theoretically based tools for investigating and analyzing the causes of human error. Therefore, they must be comprehensive too, yet they must also maintain a level of simplicity if meaningful analyses are to be performed.

Assessing comprehensiveness. So how do you know if the framework you are looking at is parsimonious, yet comprehensive enough to meet your needs today and in the future? To answer this question, one has to first decide when and where to stop looking for the causes of errors and accidents. Nearly everyone agrees that to gain a complete understanding of the causes of many accidents requires an examination well beyond the cockpit, and should include supervisors and individuals highly placed within the organization. But why stop there? As Reason (1990) and others have pointed out, there are often influences outside the organization (e.g., regulatory, economic, societal, and even cultural factors) that can affect behavior and consequently the genesis of human error. Still, if this line of reasoning is taken to its illogical extreme, the cause of every error and accident could be traced all the way back to the birth of those who erred. After all, if the individual had never been born, the error might never have occurred in the first place. Even some, in their zeal to find the "root cause," might find themselves tracing the sequence of events back to the dawn of creation. While this seems ridiculous (and it is), stopping at any point along the sequence of events is at best, a judgment call and thus is subject to the interpretation and wishes of those investigating the accident/incident.

Therefore, to circumvent the arbitrary and capricious nature of when and where to stop an investigation, many theorists have adopted the strategy of searching for "remediable causes." A remediable cause is defined as one that is readily and effectively curable, "the remedy of which will go farthest towards removing the possibility of repetition" (DeBlois, 1926, p. 48). Ideally, a comprehensive framework would be capable of capturing all the errors and their sources that, if corrected, would render the system more tolerant to subsequent encounters with conditions that produced the original error event (Reason, 1990).

With that being said, the question remains, "how do you know if the framework you are working with is truly comprehensive?" Regrettably, there are no fancy statistics or pretty diagrams to illustrate the process like there were for reliability. For all intents and purposes, it is simply a matter of mapping a given error framework onto an organization's existing accident database to see if any human causal factors remain unaccounted for. If everything is accommodated, the framework is said to be comprehensive – at least for initial analysis purposes.

All of this brings us to our initial attempt at improving the way the U.S. Navy/Marine Corps analyzes the human factors surrounding an incident/accident. Initially, we tested several "off-the-shelf" error frameworks described in the literature using the Naval aviation accident database (Shappell and Wiegmann, 1995; Wiegmann and Shappell, 1995, 1997). Much to our chagrin however, most of those frameworks focused largely on the information processing or unsafe acts level of operator performance. As a result, they did not capture several key human factors considered causal to many Naval aviation accidents.

It quickly became clear that we would need to develop a new error framework – a prospect that we did not plan for when we began our work in the area. Truth be told, our academic exercise had reached a huge roadblock until we came across James Reason's (1990) seminal work, *Human Error*. Using Reason's ideas as a starting point, we began by trying to fit the human causal factors identified within the U.S. Navy/Marine Corps accident database into the model of active and latent failures (known as the "Swiss-cheese" model of human error) described in Chapter 3. The problem was that we had to accommodate more than 700 human factors spread across more than 250 different personnel classifications (OPNAV Instruction 3750.6R).

To give you a sense of the magnitude of the task, it might be helpful to describe how the Naval Safety Center's accident database is organized. As with many accident databases, the Naval aviation database is organized around *who* committed the error, *what* occurred, and *why*. A very small part of the database has been reproduced in Tables 6.2, 6.3, and 6.4 for illustrative purposes.

Table 6.2 The person or organization involved with a given causal factor

	"Who" Factors	
Aircrew	Pilot at Controls	Mission commander
		Aircraft commander
		Instructor
		Pilot in command
		Under instruction
		Flight surgeon
		Qualified in model
		Flight leader
		Other
		No further breakdown
		Under evaluation
		Evaluator
	Pilot not at controls	Mission commander
	Table continues (see OPNAVINST 3750.6R)	

While we were able to map a number of the accident causal factors directly onto Reason's description of unsafe acts (i.e., intended and unintended behavior as described above), no pre-existing framework for the preconditions for unsafe acts and supervisory failures was available. For those two tiers we had to look for natural clusters within the U.S. Naval database. Using subject matter experts within the scientific and operational community, we were able to identify several naturally occurring categories within the database. As illustrated earlier (Figure 6.4) we identified six categories of preconditions for unsafe acts (three categories of substandard conditions and three categories of substandard practices) and seven categories of supervisory failures (three unforeseen and four known supervisory failures). Furthermore, because we used the existing Naval aviation accident database as the basis of our clustering, our newly developed taxonomy was capable of accounting for all the causal factors.

Table 6.3   What was done or not done by the individual or organization identified in Table 6.2

		"What" Factors
Aircrew	Misused flight controls on the ground	Failed to take specific necessary action
		Delayed appropriate action
		Performed wrong action
		Inadvertent operation of aircraft control
		Poor aircraft control coordination
		Misjudged speed/taxied too fast for conditions
		Misuse of nosewheel steering
		Misuse of collective and/or cyclic
		Misuse of rudders/tail rotor/tail wheel
		Misuse of nozzles
		Incorrect STO stop setting
		Misuse of brakes
		Other
	Takeoff	Improper trim setting

Table continues (see OPNAVINST 3750.6R)

Table 6.4   Why the "what" from Table 6.3 was committed

		"What" Factors
Aircrew	Communication/Coordination	Not sent
		Not received
		Received but in error
		Received but not timely
		Queue delays
		Received by not read/reviewed/implemented
		Other
	Misinterpretation – verbal	Non-standard or ambiguous language
		Incompatible verbal guidance

Table continues (see OPNAVINST 3750.6R)

Well sort of ... As described above, the *Taxonomy of Unsafe Operations* did not contain an organizational tier. This was done intentionally since very few organizational failures had been identified by U.S. Navy/Marine Corps accident investigators. For that reason, or perhaps because we were both junior officers within the U.S. Navy with career aspirations beyond our current ranks, we chose not to develop an organizational tier within our original taxonomy. However, with the landmark symposium on organizational factors and corporate culture hosted by the U.S. NTSB in 1997, it quickly became clear that organizational failures would have to be addressed within our framework if it was to be truly comprehensive. We therefore added an organizational tier in 1998, and with the other changes noted above, renamed the framework to the *Human Factors Analysis and Classification System* (Shappell and Wiegmann, 1999, 2000a; 2001). The HFACS framework was once again mapped onto the Naval aviation accident database resulting in a complete capture of the human-causal factors contributing to Naval aviation accidents (Shappell et al., 1999).

Since then, evaluations of the comprehensiveness of HFACS have been performed using data obtained from over 20,000 commercial and general aviation accidents (Wiegmann and Shappell, 2001c; Shappell and Wiegmann, 2003; in press). As you can imagine, we were delighted to find out that the HFACS framework was able to accommodate all the human causal factors associated with these accidents, suggesting that the error categories within HFACS that were originally developed for use in the military were also appropriate for civil aviation as well.

Nevertheless, while all the causal factors were accounted for, instances of some error categories within HFACS were not contained in the commercial and general aviation accident databases. As would be expected, there were very few organizational or supervisory causal factors associated with general aviation accidents. The one noted exception were accidents associated with flight training where the aircraft was owned and maintained by a flight school and the instructors acted as both an instructor and supervisor. Most other general aviation accidents involved owner/operators, so there was little need for either the supervisory or organizational tiers in those cases.

It was also interesting to note that there were no instances of organizational climate or personal readiness observed in the commercial aviation accident database, nor were there very many instances of supervisory factors. One explanation for this finding might be that contrary to Reason's model of latent and active failures, supervisory and organizational factors do not play as large a role in the etiology of commercial aviation accidents as they do in the military. On the other hand, these factors may contribute to civilian accidents, but are rarely identified using existing accident investigation processes. Based on our experience in the field, as well as those of other safety professionals within commercial aviation, the latter explanation seems more likely to be the case. Regardless, our extensive studies of HFACS' comprehensiveness indicate that it is capable of capturing the existing human causal factors contained in U.S. civil and military aviation databases.

Diagnosticity

For a framework to be effective, it must also be able to identify the interrelationships between errors and reveal previously unforeseen trends and their causes (O'Connor and Hardiman, 1996). Referred to as diagnosticity, frameworks with this quality allow analysts to identify those areas ripe for intervention, rather than relying solely on intuition and conjecture. Better yet, once trends have been identified, a truly diagnostic error framework will ensure that errors can be tracked and changes detected so that the efficacy of interventions can be monitored and assessed.

Assessing diagnosticity. Regrettably, like comprehensiveness, there are no fancy statistical measures devoted solely to the evaluation of diagnosticity. Instead, the proof is really in the proverbial pudding. Our discussion in the previous chapter of how the U.S. Navy/Marine Corps used HFACS to not only identify that violations were a problem, but also to develop an intervention strategy and then track its effectiveness, is a good example. But diagnosticity does not necessarily stop there. For a framework to be diagnostic, it must also be able to detect differences and unique patterns of errors within and among accident databases. For instance, one would expect that while there is much in common between military and civilian aviation, differences also exist, particularly when it comes to the types of operations and the way in which accidents occur. Likewise, one might expect that different types of accidents would yield unique patterns of error.

For instance, one of the most inexplicable ways to crash an aircraft is to fly a fully functioning plane into the ground. These so-called controlled flight into terrain (CFIT) accidents continue to afflict both civilian and military aviation. In fact, the U.S. Navy/Marine Corps alone lost an average of ten aircraft per year to CFIT between 1983 and 1995 (Shappell and Wiegmann, 1995, 1997b). Likewise, between 1990 and 1999, 25 percent of all fatal airline accidents and 32 percent of worldwide airline fatalities (2,111 lives lost) have been attributed to CFIT (Boeing, 2000). In fact, since 1990, no other type of accident has taken more lives within military or civilian aviation.

With this in mind, we examined 144 U.S. Navy/Marine Corps Class A accidents using an early version of HFACS (the *Taxonomy of Unsafe Operations*; Shappell and Wiegmann, 1997b). Like others working in the area, we found that many of the Naval CFIT accidents were associated with adverse physiological (e.g., spatial disorientation) and mental states (e.g., fatigue and the loss of situational awareness). In fact, to the extent that any human factors can be considered characteristic of a particular type of accident, it would be these two with CFIT.

Even more interesting however, were the differences observed in the pattern of errors associated with CFIT that occurred during the day and those that occurred at night or in the weather. As it turns out, nearly half of all CFIT accidents occur in broad daylight during VMC – a significant finding in, and of itself. After all, it had been generally felt that most, if not all, CFIT accidents occurred during the night or when visual conditions were otherwise limited, such as during IMC. Naturally then, we examined whether any differences existed in the pattern of human error associated with CFIT within these two distinctly different environments.

To no ones great surprise, there were. For instance, it is well known that when visual cues are limited, aircrew coordination, both internal and external to the cockpit, is even more critical than usual. Predictably, our analyses revealed that the incidence of CRM failures associated with CFIT was significantly higher during visually impoverished conditions. With the lack of visual cues, the proportion of adverse physiological and mental states that occurred at night or in IMC were also more prevalent than what was observed during daytime VMC.

While these findings seemed to make sense, a much more important question remained, "why would a pilot fly a perfectly good aircraft into the ground in broad daylight?" After all, to the surprise of many within the Navy hierarchy, roughly half of all CFIT occurs in daytime VMC. Again, our new error framework provided some clues as to how this could happen. It seems that many daytime CFIT accidents involved some sort of violation of the rules or regulations. What's more, these violations were often the seminal cause (if there is such a thing) in the tragic chain of events that followed regardless of whether they involved personal readiness (e.g., self-medicating or simply violating crew rest requirements) or unsafe act violations.

This latter finding was particularly important to Naval leadership since many of the interventions that had been proposed to prevent CFIT involved terrain avoidance and ground proximity warning systems (GPWS). While such technology would obviously pay dividends in preventing CFIT accidents that occur during the night or in the weather, these measures would presumably be of little help during daytime VMC – particularly, if aircrew were willing to violate established safety practices. In fact, it could be argued that the introduction of a reliable GPWS or other terrain avoidance systems might actually *increase*, rather than decrease, the likelihood that aircrew would push altitude limits in an attempt to get an edge in training or combat. Certainly, it is no stretch to envision a military pilot using an enhanced

GPWS to fly even closer to the ground in the belief that the system would bail him out if a collision was imminent.

But, CFIT is not unique to military aviation. As we mentioned above, civil aviation, in particular general aviation, has been plagued by the same problem for years. For that reason, we have recently analyzed nine years of GA accidents (1990–98) using HFACS in an effort to better understand the cause of CFIT in civilian operations (Shappell and Wiegmann, 2003).

Our analysis included over 14,000 GA accidents, of which roughly 10 percent (1,407) were classified as CFIT by our pilot-raters. Consistent with what we saw within the U.S. Navy/Marine Corps, almost one-third of all CFIT accidents were associated with violations of the rules – nearly three times more than what was seen with non-CFIT accidents (Table 6.5). Likewise, personal readiness failures, arguably another type of violation that occurs external to the cockpit, were over four times more likely during CFIT. As expected, adverse mental states and perceptual errors[14] were also more prevalent during CFIT than non-CFIT accidents.

Table 6.5 **CFIT and non-CFIT accidents associated with at least one instance of a particular causal category**

	CFIT [a]		Non-CFIT	
	Total	**%**	**Total**	**%**
Unsafe Acts of Operators				
Errors				
Decision Errors	471	33.5	4472	35.3
Skill-based Errors	1074	76.3	9286	73.2
Perceptual Errors	176	12.5	911	7.2
Violations [b]	445	31.6	1574	12.4
Preconditions for Unsafe Acts				
Substandard Conditions of Operators				
Adverse Mental States	171	12.2	576	4.5
Adverse Physiological States	52	3.7	317	2.5
Physical/Mental Limitations	182	12.9	2392	18.9
Substandard Practices of Operators				
Crew Resource Management	102	7.2	1397	11.0
Personal Readiness	89	6.3	206	1.6

[a] Significant differences between CFIT and non-CFIT (p<.001) are shaded.
[b] Violations have been collapsed across routine and exceptional violations.

---

[14] The *Taxonomy of Unsafe Operations* used with the U.S. Navy/Marine Corps data did not include a perceptual errors category. Retrospectively, many of the adverse physiological states seen in the Naval aviation data would be coded as perceptual errors using HFACS.

Given our findings from the U.S. Navy/Marine Corps study, it seemed reasonable to explore whether similar differences existed between GA CFIT accidents occurring in broad daylight and those that occurred in visually impoverished conditions. Like the military data, GA CFIT accidents were evenly split between those that occurred in clear daytime conditions and those that occurred at night, or in IMC. Those that occurred during visually impoverished conditions were often associated with adverse physiological states, physical/mental limitations, and poor CRM (Table 6.6). Furthermore, CFIT accidents were six times more likely to involve a violation of the rules if they occurred in visually impoverished conditions. Indeed, it is not hard to envision a crew that fails to obtain a weather update prior to takeoff (crew resource management) and consequently encounters adverse weather enroute. Then, after choosing to continue into IMC while VFR only (violation), they end up spatially disoriented (adverse physiological state) and collide with the terrain.

Table 6.6 CFIT accidents occurring in clear versus visually impoverished conditions

	Clear [a]		Impoverished	
	Total	%	Total	%
Unsafe Acts of Operators				
Errors				
Decision Errors	216	31.5	241	34.7
Skill-based Errors	573	83.6	480	69.1
Perceptual Errors	101	14.7	74	10.6
Violations [b]	91	13.3	346	49.8
Preconditions for Unsafe Acts				
Substandard Conditions of Operators				
Adverse Mental States	66	9.6	103	14.8
Adverse Physiological States	9	1.3	40	5.8
Physical/Mental Limitations	65	9.5	115	16.5
Substandard Practices of Operators				
Crew Resource Management	20	2.9	82	11.8
Personal Readiness	36	5.3	53	7.6

[a] Significant differences between visual and impoverished visual conditions ($p<.001$) are shaded.
[b] Violations have been collapsed across routine and exceptional violations.

As easy as it may be to rationalize how GA pilots could fly their aircraft into the ground at night or in IMC, trying to comprehend why a pilot would collide with terrain in clear daytime conditions is considerably more perplexing. Unlike what we have seen within the Naval aviation database, it does not seem to be simply a function of an overconfident pilot pushing the envelope to gain an advantage over an adversary or target. No, violations are considerably less prevalent among GA CFIT accidents that occur in clear daytime conditions. The only variable that was more prevalent was skill-based errors, suggesting that CFIT in daytime VMC may be the result of inattention, a breakdown in visual scan, distraction, or simply stick-and-rudder skills. The fact that one half of all GA CFIT occur in these seemingly benign conditions suggest that further work is required to fully understand this type of accident.

In a sense then, HFACS has provided us a unique look into the causes of CFIT and has illuminated (diagnosed) potential areas in need of intervention. However, often lost in the discussion, is the benefit of a common error framework when it comes to identifying trends between quantifiably different databases as unique as those within military and civilian aviation sectors. In fact, HFACS has now been implemented within all four branches of the U.S. military, as well as both commercial and general aviation. As a result, we can now compare error trends among these different organizations. But, will HFACS be diagnostic enough to detect any differences that would be expected to exist and even some that we had not anticipated? Let us take a look at a couple of error categories to find out.

The percentage of U.S. military and civil aviation accidents (1990–98) associated with at least one perceptual error is presented in Figure 6.7. As expected, a larger percentage of military accidents involve perceptual errors than do civil aviation accidents. This is not surprising given that military pilots typically engage in more dynamic and aggressive flight, resulting in unusual attitudes that often wreak havoc with one's vestibular system causing spatial disorientation and perceptual errors. Even military helicopter pilots periodically perform low-level flights at relatively high speeds at night while donning night vision goggles, all of which increase the opportunity for perceptual errors.

Such findings have implications for the development of intervention strategies, as well as for gauging the amount of time and resources that should be invested in addressing perceptual errors in aviation. Obviously for military aviation, this issue will require a great deal more attention since preventing accidents due to perceptual errors could substantially improve safety. Within civil aviation, however, considerably fewer resources should likely be invested, given that the problem of perceptual errors is relatively small. Albeit, preventing all types of accidents is important, but when resources are limited and there are bigger problems to address, one must be judicious, ensuring that the problems having the biggest impact on safety are given priority.

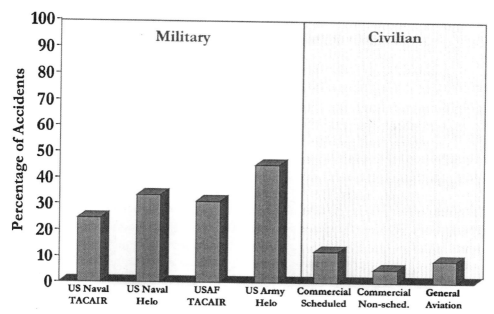

Figure 6.7 Percentage of accidents associated with perceptual errors across military and civilian aviation (1990–98)

But what is the "big" problem? Well, many have suggested that it is decision-making or even CRM. Indeed, as we mentioned in the previous chapter, a great deal of time and money has already been invested in developing programs to address these issues, based on the assumption that these are the major causes of accidents. However, the results from our HFACS analysis actually suggest that rather than decision-making or CRM, skill-based errors are associated with the largest percentage of aviation accidents – particularly, but not exclusively, within the civil aviation sector.

As illustrated in Figure 6.8, more than half of all civil aviation accidents are associated with skill-based errors. The observation that the largest percentage was within general aviation is not surprising given the amount of flight time, training, and the sophistication of the aircraft within the GA community relative to the other aviation domains. While it can certainly be argued that there are exceptions, most would agree that the average GA pilot does not receive the same degree of recurrent training and annual flight hours that the typical commercial or military pilot does. Likewise, we would all agree that most GA aircraft are not nearly as sophisticated as an F-14 Tomcat or Boeing 777.

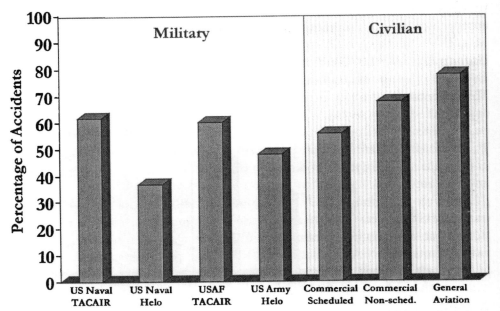

Figure 6.8 Percentage of accidents associated with skill-based errors across military and civilian aviation (1990–98)

Given the nature of non-scheduled air carrier operations and the relative experience of the aircrew, it was somewhat expected that the next highest percentage of accidents associated with skill-based errors was within the category of on-demand, non-scheduled air carriers. This is not to say that these so-called "air-taxi" pilots are unskilled or inexperienced, just that relative to their counterparts within the military and major air-carriers, they generally have less training and experience. In many instances, the aircraft they fly are less sophisticated as well.

What is perhaps more interesting, and even somewhat surprising, is that nearly 60 percent of all accidents involving scheduled air carriers were associated with at least one skill-based error. Apparently, even the most highly trained, experienced, and skilled pilots continue to have difficulty with such issues as managing their attention and memory resources as well as handling the workload required to fly modern commercial aircraft. Even more surprising is the finding that commercial pilots were nearly as susceptible to skill-based errors as military pilots flying tactical fighter aircraft were. While we are only now exploring this issue, there is clearly much to be learned from findings such as these.

In wrapping up this section, we want to reiterate that there really is no good measure of a framework's diagnosticity. That being said, we have made an earnest attempt to illustrate how it might be inferred based on research findings like the ones presented here. Indeed, others may have a different approach or method, but they will have to write their own book...

Usability

The acceptance of an error analysis approach is often determined by how easy it is to use. In reality, it matters very little whether a system is reliable, comprehensive, or diagnostic if investigators and analysts never use it. Therefore, the usability of a framework, or the ease with which it can be turned into a practical methodology within an operational setting, cannot be ignored (Hollnagel, 1998).

Since its inception, the acceptability of HFACS and its predecessor frameworks has been repeatedly assessed and improved, based on feedback from those attending our training sessions, as well as from operators who have been using HFACS in the field. Over the years, input has come from pilots, flight surgeons, aviation safety officers, and other safety personnel within both military and non-military organizations from around the world.

Some changes that we have made to improve the acceptability and usability of HFACS have included the rephrasing of technical or psychological terminology (e.g., slips, lapses and mistakes), to create terms that aviators would better understand (e.g. skill-based and decision errors). Other changes in the nomenclature were also made. For example, our use of the term taxonomy often drew blank and/or quizzical stares, as our students often thought we were going to teach them how to stuff animals. By changing the name of our error framework from the *Taxonomy of Unsafe Operations* to the *Human Factors Analysis and Classification System* or HFACS, we not only made the system more palatable to potential users, but we were no longer confused with taxidermists!

The clearest evidence of HFACS' usability however, is that large organizations like the U.S. Navy/Marine Corps and the U.S. Army have adopted HFACS as an accident investigation and data analysis tool. In addition, HFACS is currently being utilized within other organizations such as the FAA and NASA as a supplement to their preexisting systems (Ford et al., 1999).

Perhaps the main reason why these organizations have embraced HFACS is that the system is highly malleable and adaptable. In other words, HFACS can be easily modified to accommodate the particular needs of an organization. For example, the role that environmental factors play in the etiology of human error was not incorporated into the version of HFACS developed for the U.S. Navy/Marine Corps, primarily because factors such as weather or terrain are uncontrollable and hence not considered remedial causes of an accident by the U.S. Naval Safety Center. Still, in some contexts, environmental factors may be controllable, such as in a maintenance facility or an air traffic control center. Therefore, these organizations may view the role of environmental factors differently or weigh them more heavily. In addition, some have argued that working conditions and equipment design problems are not only an organizational resource management issue, but also a precondition of unsafe acts. Therefore, such factors need to be addressed separately during the analysis of an accident. Given all of this, the version of HFACS that we have presented in this book includes these additional environmental and technological categories for those organizations that may find them useful.

Additional evidence of HFACS' adaptability comes from organizations that have taken the liberty of modifying the framework themselves, successfully tailoring it to suit their own needs. For example, the Canadian Forces has developed what they call CF-HFACS, which is presented in Figure 6.9. As a minor modification to its depiction, they have chosen to place the unsafe acts tier at the top and organizational influences at the bottom in order to represent the typical analysis process that often starts with the identification of an unsafe act. They have also chosen to remove the negative wording of the categories, perhaps to reduce the appearance of apportioning blame to those individuals whose actions were causal to the accident.

Less cosmetic, is the addition of categories at both the unsafe acts and preconditions level. As can be seen, they have chosen to separate technique errors from skill-based errors, which are now labeled "attention/memory." Also added is a "knowledge information" category, which is a type of error that occurs when knowledge or the information available to complete a task is incorrect, impractical or absent. Other changes at the pre-conditions level include the addition of categories such as "qualifications" and "training." Although this information is already included in HFACS, creating separate "bins" for such factors highlights their importance placed on them by this organization.

Other variations of the original HFACS framework exist. For example, there is a version of HFACS for air traffic control (HFACS-ATC), aircraft maintenance (HFACS-ME), and even one for medicine (HFACS-MD). Descriptions of all of these frameworks can be found in the open literature (Pounds et al., 2000; Schmidt et al., 1998; Wiegmann et al., in press). Although on the surface, each of these derivatives may appear different from the parent HFACS, they all have HFACS as their core. Changes that were made to specific category descriptors were done primarily to accommodate the idiosyncrasies of their target audience. For example, since air traffic controllers tend to bristle at the term "violations," HFACS-ATC has replaced the term with the word "contraventions." Although anyone with a thesaurus can tell you, "contraventions" means the same thing as violations, the term is apparently much more palatable to its users. Perhaps this is because no one actually knows what contravention really means! However, as Shakespeare once wrote, "a rose by any other name is still a rose." But seriously, we welcome and even encourage others to modify HFACS to meet their needs. Safety is much more important to us than protecting our egos. Still, we hope that all who use or modify the system will let us know, because we often learn from it and actually appreciate it when others do our work for us!

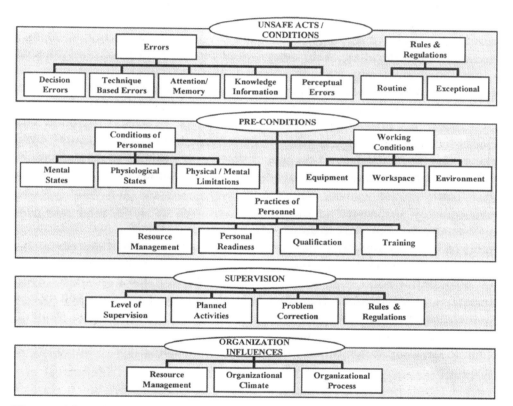

Figure 6.9   HFACS as modified by the Canadian Forces (CF-HFACS)

Conclusion

First of all, let us congratulate you for making it through this entire chapter. Hopefully you did not get so bored that you jumped right to this summary section. But as we warned you from the outset, this chapter is by far the most academic in the book. Our purpose in writing it was to provide an overview of the criteria that can be used for evaluating error frameworks including HFACS. Unfortunately, some may view this chapter as simply self-serving or a way of touting HFACS as being the best. Please, believe us, this was *not* our intention. Rather, we have taken the opportunity to use our firsthand knowledge of HFACS' development to help describe these criteria and illustrate the evaluation process. We also wanted to provide the requisite information that potential users of HFACS will need to objectively evaluate HFACS for themselves.

Admittedly, we are not the first to propose evaluation criteria or use them in the development of an error analysis system. However, all too often we have witnessed people haphazardly put together frameworks and then market them as being the "state of the art" without any validation whatsoever. As a result, the users of such systems are left to their own intuition or "gut feelings" when deciding on the adequacy of a given error framework. Indeed, often times such systems are accepted as valid simply based upon the reputation of the individual or organization who developed it.

Hopefully, our readers will now be able to get beyond gut feelings when evaluating an error framework by asking its developer very pointed questions, such as: "What is the inter-rater reliability of your system?", "Will the system accommodate all the human factors information in my database?", "Will the framework allow me to identify unique trends in my error data so that interventions can be targeted at key problems and their efficacy subsequently evaluated?", "Can your framework be understood and used universally by our investigators and safety personnel?", and "Can it be tailored to meet my organization's specific needs?". These are but a few questions that an educated consumer should ask when shopping around for an error framework. If its developer cannot answer these or other related questions to your satisfaction, then you should move on until you find an approach that meets your expectations.

Speaking of moving on, however, that is what we plan to do now. It is time to catch up to the others who skipped this chapter and went straight to chapter 7.

7 But What About...?

We would love to say that everyone who has ever heard or read about HFACS has immediately become an unwavering believer in the system. But, like any framework or method of accident investigation, HFACS too has its critics. There have been several occasions during our workshops or conference presentations, or while reviewing one of our manuscripts, that others in the field have voiced their questions or concerns. Some of these critiques have been due to differences in basic philosophical views concerning the fundamental nature of human error and accident causation. While such debates have helped refine our thinking, they have often led to an unavoidable impasse, ultimately boiling down to one's belief as to whether or not human error really even exists at all. Others, however, have offered very insightful critiques that have led to significant improvements in the present version of HFACS. We have always tried not to take any of these comments and critiques personally. Yet, as any proud parent gets upset when others criticize their children, we too have had our moments. Nevertheless, if any framework of accident investigation is to live beyond the personalities of those who created it, it must be flexible yet robust enough to withstand concerns and questions presented by critics.

Admittedly, HFACS is not perfect and our academic training and consciences compel us to be "honest brokers" in our presentation of the system. After all, no one chooses a career in safety if they want to make a lot of money. In fact, HFACS has been published in the open scientific literature and can be used by anyone, free of charge. We do not make any money from "selling it." As anyone who has ever published a textbook knows, you do not get much in royalties either. Albeit, we might earn enough to buy ourselves a cup of coffee, but we're certainly not going to be able to fund our children's college education from the sale of this book! So, in all fairness to those who have offered their critiques of HFACS, we have decided to dedicate this last chapter to the systematic presentation of the comments and criticisms we have encountered over the years. After all, if one person has raised the question, there may be others with similar concerns.

We have named this chapter "But What About...?" because it represents a type of "frequently asked questions" or FAQ document that might be found in an HFACS user's guide. We have chosen to state these questions and concerns directly, sometimes phrasing them in the rather blunt fashion that we originally received them, and then to briefly address each one in turn. Our hope is that in reading this chapter you too will have several of your questions or concerns answered and that ultimately you will be better able to determine the potential utility of HFACS for your organization.

Isn't there more to human error than just labels?

Some critics of HFACS have argued that the framework simply relies on labels such as "decision errors" or "crew resource mismanagement" to give investigators access to the psychological life beneath an error. What is needed is more guidance beyond the cursory analysis presented here, if one is to truly understand the genesis of human error.

From an academic perspective, this is indeed a legitimate concern. However, we have intentionally restrained ourselves from writing an academic treatise on human error, as several titles already exist including James Reason's (1990) book "Human Error," upon which HFACS is based. Rather, we have chosen to develop a framework and to write a book that is aimed at practitioners. Our intent was not to describe all the theoretical underpinnings associated with each of the categories within HFACS. The end result would have taken on encyclopedic form. Rather, our goal was to provide a much-needed, rudimentary tool for applying a systematic human error approach to accident investigation. As such, HFACS provides a "down-to-earth," practical framework for helping investigators identify the *need* to further investigate the possible error forms that potentially contributed to an accident. Information for conducting these additional analyses can be readily found in the literature or by consulting experts in a particular area.

Frameworks or checklists only limit the scope of the investigation.

While we appreciate the possible drawbacks of using checklists during accident investigations, we do not view HFACS as a framework that confines investigators, or encourages them to consider only a limited number of possibilities. Rather, HFACS suggests multiple avenues of inquiry when analyzing human error and encourages investigators to go beyond what a pilot may have done wrong to discovering why the error occurred. Indeed, our discussions with those who actually use HFACS in the field indicate that HFACS actually helps them expand their investigation beyond just the facts about an accident. As we have stated previously in other chapters, most investigators are not formally trained in human factors and therefore they have often stopped at the level of "pilot error" when investigating accidents. Consequently, without frameworks like HFACS, the majority of investigations have historically left large gaps in the information collected (such as information about preconditions, and supervisory and organizational issues). HFACS has therefore actually *increased* both the quantity and quality of the human factors data collected during aviation accident investigations.

What is really missing in HFACS is an emphasis on error-producing factors related to equipment design.

This is by far the most common criticism that we have received about HFACS over the years. Indeed, our initial focus was, and continues to be, on those responsible for the design, specification, and procurement of equipment rather than on the equipment itself, since only humans can effect change. For example, if specifications for equipment were poor or if the organization chose to go a cheaper route and purchased sub-standard equipment, these causal factors would be examples of organizational failures, specifically those involving "resource management" using HFACS. In essence then, our focus is on the human aspects, not necessarily engineering or "knobs and dials" approaches. Nonetheless, we have listened to our critics and expanded HFACS in this book to include error producing factors such as equipment design and other environmental factors to address some of the issues that admittedly do not fall completely under the organizational umbrella. However, we remain convinced that unless there is some form of human responsibility attached to such problems, they will never get changed within the system. A poor computer interface cannot change itself no matter how many times you cite it as a cause to an accident.

HFACS oversimplifies the translation from "finding a hole" to "plugging it up" and these are definitely not the same thing.

We agree that there is much more to accident prevention and mitigation than HFACS. In fact, HFACS is only the first step in the risk management process – i.e., the identification of the problem. Too often, organizations embark on intervention strategies based on anecdotes and gut feelings rather than objective data that indicates what the problem really is. As we have often pointed out, many organizations subscribe to the "mishap-of-the-month" club. That is, if a particularly noteworthy accident involves CRM failures, all of the sudden millions of dollars are spent on CRM training. Then next month, it may be an accident involving violations of the rules. So, they move the dollars to procedural enforcement strategies and accountability only to find out that next month they have a controlled flight into terrain accident and the dollars are once again moved to prevent or mitigate causal factors associated with these types of accidents. In essence, these organizations are left doing little more than chasing their tails. However, the HFACS framework has proven useful to the U.S. Navy/Marine Corps and other organizations in providing data-driven efforts based on a series of accidents/incidents, rather than isolated cases that may have little to do with the more pervasive problems within the system. The approach also provides a way of monitoring the effectiveness of interventions so that they can be revamped or reinforced to improve safety.

The analysis of supervisory failures in HFACS has a large problem of hindsight. After the fact, it is always easy to identify where supervisors or management should have paid more attention.

Such a statement implies that the criterion for determining whether a factor was causal to an accident is whether it was identifiable prior to its occurrence. This seems a little silly and would likely result in a great deal of human and engineering data to be dismissed prematurely during an accident investigation. Such a statement also appears to be based more on a concern about the issue of blame rather than accident prevention (we'll address the blame issue later in this chapter). Furthermore, it is no big secret, and therefore no major revelation, that all accident investigations involve "hindsight" to some extent. Yet, to simply ignore data because it was discovered "after the fact" would be nothing short of malpractice. After all, much of the learning process involves changing behavior once an error has occurred. From a safety perspective, it is not criminal to make an error, but is inexcusable if you don't learn from it.

We designed HFACS as a tool to help investigators ask the right questions, so that supervisory and organizational issues are at least considered. HFACS also provides a tool to examine and track trends in human error and accident data. As a result, intervention strategies can be developed that are not based on hindsight, anecdotes or "shoot from the hip" ideas, but rather on specified needs and objective data. Furthermore, researchers at the FAA's Civil Aerospace Medical Institute recently reported on a safety survey that they developed based on HFACS (Bailey et al., 2000). The survey was used with regional air carriers in Alaska to identify those human error areas that discriminated between carriers that had experienced an accident in the last three years and those that did not. This initial effort to develop a proactive survey to identify hazardous supervisory issues before they cause accidents, has shown considerable promise.

It is not difficult to develop a human factors evaluation framework that utilizes under-specified labels and subsequently achieve high inter-rater reliabilities.

From our experience, this comment does not appear to be true. In fact, we have found quite the opposite regarding "under-specified labels" and inter-rater reliabilities. As a matter of fact, until we refined the causal categories associated with the "Taxonomy of Unsafe Operations" (as the HFACS framework was originally called), and provided crisp definitions that were mutually exclusive, our inter-rater reliabilities were not nearly as high as those we enjoy today. This is exactly where many error analysis frameworks fall short. Indeed, other authors have presented similar systems, professing them to be user friendly and reliable without any appreciable data to back it up. This is perhaps one of HFACS' strongest suits. That is, HFACS is based on several years of extensive testing using thousands of military and civil aviation accidents and consequently there is considerable empirical data supporting its utility.

When it comes to evaluating a framework like HFACS, it may be more meaningful to compare it to other similar tools.

Another good point, however, we really do not want to engage in any downplaying of other frameworks directly. Rather, our aim was to put some evaluation criteria on the table and let the reader do his or her own comparisons. However, the fact that several organizations around the world have compared HFACS to their own systems and have chosen to adopt HFACS is clear evidence of its utility. Nevertheless, there are often redeeming qualities associated with almost all error frameworks. As a result, some organizations like the FAA and EUROCONTROL have chosen to merge HFACS with other systems (like the Human Error Reliability Analysis (HERA)), to capitalize on the benefits of both systems when investigating and analyzing ATC operational errors. Still others have chosen to keep their existing systems, and have employed HFACS as an adjunct to their pre-existing systems. Regardless of how it is employed, our hope is that those who are in need of a better human error system will find HFACS helpful in improving safety within their organizations.

If I adopt HFACS, won't I lose all the accident data that I already collected with my old system?

Since the HFACS framework is causal-based (e.g., skill-based error) rather than descriptive-based (e.g., failed to lower landing gear), it is "transparent" to the original causal factor identification in most databases. That is, application of the HFACS framework does not require a reinvestigation of the accident or changes to be made in the original accident causal factors; they can simply be regrouped into more manageable categories. This means that existing accidents can be classified post hoc, allowing for analysis of trends now, rather than years down the road, when sufficient numbers of accidents are investigated using a new framework.

Implicit in the principle of a transparent framework is that the integrity of the existing database is preserved. After all, existing database formats have proven particularly useful in the management and reduction of mechanical failures. To simply replace existing database structures with new human factors databases would be like "throwing the baby out with the bath water." By superimposing the HFACS framework onto existing databases, you can preserve the benefits realized by existing frameworks while adding the obvious advantages of a human factors database. This is exactly what organizations like the U.S. Navy/Marine Corps have done.

Numerous theories of accident causation have come and gone over the years. Therefore, since HFACS is tied to Reason's theory of latent and active failures, it is as fad-based as the analysis tools the authors wanted to get away from.

James Reason's 1990 book "Human Error" started a revolution of sorts that breathed new life into accident and incident investigation, both in the fields of nuclear power and aviation. HFACS capitalizes on this work by providing a framework for applying Reason's ideas and theory. The fact that the book "Human Error" was first published over a decade ago and continues to be one of the most widely cited and respected works in this area speaks for itself.

Isn't HFACS just the same old "blame-and-train" game?

We chose to end our book addressing this question because it reflects what we believe is a growing chasm between two major schools of thought within the field of human factors. Furthermore, where one falls on either side of this chasm often determines whether HFACS is embraced or discarded. Therefore, we will spend a little more time elaborating upon this issue and then discuss what we think it really means for the utilization of HFACS.

Consider the following scenarios: An automobile driver gets into an accident after running a stoplight and blames it on the yellow caution light, which varies in duration across signals in the community. A shopper hurts his back while loading groceries into the trunk of a car and blames it on the bags, which were designed to hold too many groceries. A student over-sleeps and misses an exam then blames it on his alarm clock that failed to go off at the pre-programmed time. A citizen fails to pay income taxes and blames it on the forms, which are too complex to understand. Such anecdotes illustrate the dilemma we often face as human factors professionals in attributing causes to accidents and errors in the workplace. But how do we (or should we) view operator error in light of this dilemma?

As discussed in Chapter 2, some theories of human error focus on expanding models of human information processing to help explain the types of errors commonly made by operators in different contexts. These theories have helped raise awareness of human limitations, in terms of attention and cognitive processing abilities, and have encouraged a more thorough analysis of human error during accident investigations. Although these approaches have contributed greatly to our understanding of human error, an unwanted side-effect can emerge, which is the tendency to assign blame to operators directly involved in the unsafe act, particularly by those who have vested interests in the outcome of such analyses. This can lead to a "blame-and-train" mentality. Unfortunately, however, such adverse consequences have been generalized to this entire error analysis approach, so that some refer to all such models as "bad apple" theories of human performance.

Another common approach to human error analysis is the systems or ergonomics approach. This is an "alternate" view of human error that considers the interaction between human and technical factors when exploring the causes of errors and accidents. Indeed, most students of human factors are likely taught this approach. There is no doubt that this approach has helped highlight the affects that system design can have on operator performance. It has also led to a greater appreciation of design-induced errors and an understanding of how to create systems that are more error-tolerant. However, as mentioned in Chapter 2, this approach often fails to consider intricate aspects of the human side of the error, relative to the information processing or cognitive approaches. Furthermore, it has led some in the field of human factors to conclude that human error does not really exist. Rather, when an accident occurs, it is the "system" that has erred, not the human.

Although many human factors professionals appreciate the strengths and weakness of these various approaches, we must ask whether some in the field have taken the concept of "designed-induced" error too far? Indeed, there does appear to be some who hold the opinion that operators are simply "victims" of inadequate design – a type of "the designer made me do it" mentality. Even more disturbing, however, is that such extremism appears to be becoming increasingly more pervasive. But, is this perspective really correct? Should we never attribute at least some responsibility to operators who err?

Consider for example an airplane that crashes while taking off down a closed runway and the surviving aircrew blame it on the poor runway signage and taxi lights. But then how should we interpret that fact that aircrew of 10 other aircraft successfully navigated the same route just before the accident occurred and were able to take-off on the correct runway? A more personal example occurred to a colleague who was helping review papers for a professional human factors conference. An author who was late submitting a proposal for the conference contacted him, blaming the tardiness of his paper on the instructions, which were too confusing. When the reviewer replied that 20 other presenters had interpreted the instructions correctly and had mailed their proposals in on time, the author stated that the reviewer was "blaming the victim" and that his "bad attitude was surprising for someone in human factors."

Was it a really a bad attitude, or was the reviewer too just a "victim" of the constraints imposed on him by the organization for which he was reviewing proposals? Seriously, how should we view all those internal performance-shaping factors outlined in HFACS, such as operator expectancies, biases, complacency, over-confidence, fatigue, technique, time-management, aptitude, attitudes, and personality? Do these not count? When an error does occur, can we say only that the engineers failed to design a system that was adaptable to all levels of size, shape, knowledge, skill, and experience of operators? Is taking the time to read instructions, ask questions, plan activities, communicate intentions, and coordinate actions no longer the responsibility of operators? If such things are considered in analyzing operator error, is one practicing "bad" human factors? Obviously, we don't think so, but there are some who would disagree. They argue that we are using HFACS to push the pendulum too far back toward the "blame-and-train" direction. However, we believe that we are simply pushing it back toward middle ground, where both equipment design *and* operator characteristics are considered together. But then again, maybe we are just a couple of bad apples who are in need of some good company!

References

Adams, E. (1976, October). Accident causation and the management system. *Professional Safety, 21*(10), 26–29.

Anastasi, A. (1988). *Psychological testing* (6th ed.). New York: Macmillan.

Bailey, L., Peterson, L., Williams, K., and Thompson, R. (2000). *Controlled flight into terrain: A study of pilot perspectives in Alaska* (DOT/FAA/AM-00/28). Washington, DC: Federal Aviation Administration, Office of Aerospace Medicine.

Bird, F. (1974). *Management guide to loss control*. Atlanta, GA: Institute Press.

Boeing (2000). *Statistical summary of commercial jet airplane accidents: Worldwide operations 1959-1999*. [On-line]. Available: www.boeing.com/news/techissues/pdf/1999_statsum.pdf.

Brenner, C. (1964). Parapraxes and wit. In W. Haddon, Jr., E.A. Suchman and D. Klein (Eds.), *Accident research: Methods and approaches* (pp. 292–295). New York: Harper and Row.

Cacciabue, P.C. (2001). Human factors insight and data from accident reports: The case of ADREP-2000 for aviation safety assessment. In D. Harris (Ed.) *Engineering psychology and cognitive ergonomics, Vol. 5. Aerospace and transportation systems* (pp. 315–324). Burlington, VT: Ashgate Publishing.

Carmines, E.G. and Zeller, R.A. (1979). *Reliability and validity assessment*. Thousands Oaks, CA: Sage Publications.

DeBlois, L. (1926). *Industrial safety organization for executive and engineer*. London: McGraw Hill.

Degani, A. and Wiener, E.L. (1994). Philosophy, policies, procedures, and practice: The four "P's" of flight deck operations. *Aviation psychology in practice* (pp. 44–67). Brookfield, VT: Ashgate.

Diehl, A. (1992, January). Does cockpit management training reduce aircrew error? *ISASI Forum, 24*.

Edwards, E. (1988). Introductory overview. In E. Wiener and D. Nagel (Eds.), *Human factors in aviation* (pp. 3–25). San Diego, CA: Academic Press.

Federal Aviation Administration (1997). *Crew resource management training* (Report Number AC 120-51B). Washington DC: Author.

Ferry, T. (1988). *Modern accident investigation and analysis* (2nd ed.). New York, NY: Wiley and Sons.

Firenze, R. (1971, August). Hazard control. *National Safety News, 104*(2), 39–42.

Fleiss, J. (1981). *Statistical methods for rates and proportions*. New York: Wiley.

Flight Safety Foundation, (1997). Graph of projected traffic growth and accident frequency. Retrieved September 02, 1997, from http://www.flightsafety.org.

Ford, C.N., Jack, T.D., Crisp, V. and Sandusky, R. (1999). Aviation accident causal analysis. In *Advances in Aviation Safety Conference Proceedings* (p. 343). Warrendale, PA: Society of Automotive Engineers Inc.

Fraser, J.R. (2002). Naval aviation safety: Year in review. *Aviation Space and Environmental Medicine, 73*(3), 252.

Fuller, R. (1997). Behaviour analysis and aviation safety. In N. Johnston, N. McDonald, and R. Fuller (Eds.), *Aviation psychology in practice* (pp. 173–189). Brookfield, VT: Ashgate.

Government Accounting Office (1997). *Human factors: FAA's guidance and oversight of pilot crew resource management training can be improved* (GAO/RCED-98-7). Washington, DC: Author.

Haddon, W., Jr., Suchman, E.A. and Klein, D. (1964). *Accident research: Methods and approaches.* New York: Harper and Row.

Heinrich, H.W., Petersen, D. and Roos, N. (1980). *Industrial accident prevention: A safety management approach* (5th ed.). New York: McGraw-Hill.

Helmreich, R.L. and Foushee, H.C. (1993). Why crew resource management? Empirical and theoretical bases of human factors training in aviation. In E.L. Wiener, B.G. Kanki and R.L. Helmreich (Eds.), *Cockpit resource management* (pp. 3–45). San Diego, CA: Academic Press.

Hollnagel, E. (1998). *Cognitive reliability and error analysis method (CREAM).* Oxford: Alden Group.

International Civil Aviation Organization (1993). *Investigation of human factors in accidents and incidents* (Human Factors Digest #7). Montreal, Canada: Author.

Jensen, R.S. and Benel, R.A. (1977). *Judgment evaluation and instruction in civil pilot training* (Final Report FAA-RD-78-24). Springfield, VA: National Technical Information Service.

Johnson, W. (1997). *Classifying pilot human factor causes in A-10 class A mishaps.* Unpublished Graduate Research Project. Daytona Beach, FL: Embry-Riddle Aeronautical University.

Jones, A.P. (1988). Climate and measurement of consensus: A discussion of "organizational climate". In S.G. Cole, R.G. Demaree and W. Curtis (Eds.), *Applications of interactionist psychology: Essays in honor of Saul B. Sells* (pp. 283–290). Hillsdale, NJ: Erlbaum.

Kayten, P.J. (1993). The accident investigator's perspective. In E.L. Wiener, B.G. Kanki and R.L. Helmreich (Eds.), *Cockpit resource management* (pp. 283–314). San Diego, CA: Academic Press.

Kern, T. (2001). *Controlling pilot error: Culture, environment, and CRM.* New York: McGraw Hill.

Lahm, F.P. (1908, September). R*eport of the accident to the Wright Aeroplane at Ft. Myer, Virginia, on September 17, 1908.* Retrieved February 10, 2003, from http://www.arlingtoncemetery.com/thomaset.htm.

Lauber, J. (1996). Forward. In R. Reinhart (Author). *Basic flight physiology* (2nd ed., pp. xi). New York: McGraw-Hill.

Lautman, L. and Gallimore, P. (1987, April-June). Control of the crew caused accident: Results of a 12-operator survey. Airliner, 1–6. Seattle: Boeing Commercial Airplane Co.

Muchinsky, P.M. (1997). *Psychology applied to work* (5th ed.). Pacific Grove, CA: Brooks/Cole Publishing Co.

Murray, S.R. (1997). Deliberate decision making by aircraft pilots: A simple reminder to avoid decision making under panic. *The International Journal of Aviation Psychology, 7*(1), 83–100.

Nagel, D. (1988). Human error in aviation operations. In E. Wiener and D. Nagel (Eds.), *Human factors in aviation* (pp. 263–303). San Diego, CA: Academic Press.

National Transportation Safety Board. (1973). *Eastern Air Lines, Inc., L-1011, N310EA, Miami, Florida, December 29, 1972* (Technical Report NTSB-AAR-73-14). Washington: National Transportation Safety Board.

National Transportation Safety Board. (1993). *Tomy International, Inc. d/b/a Scenic Air Tours, Flight 22, Beech E18S, N342E, Maui, Hawaii, April 22, 1992* (Technical Report NTSB/ARR-93/01). Washington, DC: Author.

National Transportation Safety Board (1994a). A review of flightcrew-involved, major accidents of U.S. Air Carriers, 1978-1990. (Safety Study PB94-917001), Washington, DC: Author.

National Transportation Safety Board (1994b). *Uncontrolled collision with terrain. American International Airways flight 808, Douglas DC-8-61, N814CK, U.S. Naval Air Station, Guantanamo Bay, Cuba, August 18, 1993* (Report Number NTSB/AAR-94-04). Washington, DC: Author.

National Transportation Safety Board. (1995a). *Transportes Aereos Ejecutivos, S.A. (TAESA), Learlet 25D, XA-BBA, Chantilly, Virginia, June 18, 1994* (Technical Report NTSB/ARR-95/02). Washington, DC: Author.

National Transportation Safety Board. (1995b). *Air Transport International, Douglas DC-8-63, N782AL, Kansas City, Missouri, February 16, 1995* (Technical Report NTSB/ARR-95/06). Washington, DC: Author.

Nicogossian, A.E., Huntoon, C.L., and Pool, S.L. (1994). *Space physiology and medicine* (3rd ed.). Baltimore, MD: Williams and Wilkins.

O'Connor, S.L. and Hardiman, T. (1996). Human error in civil aircraft maintenance and dispatch: The basis of an error taxonomy. *Paper Presented at the First International Conference on Engineering Psychology and Cognitive Ergonomics*.

O'Hare, D., Wiggins, M., Batt, R. and Morrison, D. (1994). Cognitive failure analysis for aircraft accident investigation. *Ergonomics, 37*(11), 1855–1869.

Orasanu, J.M. (1993). Decision-making in the cockpit. In E.L. Wiener, B.G. Kanki and R.L. Helmreich (Eds.), *Cockpit resource management* (pp. 137–172). San Diego, CA: Academic Press.

Peterson, D. (1971). *Techniques of safety management*. New York: McGraw-Hill.

Plourde, G. (1997). *Human factor causes in fighter-bomber mishaps: A validation of the taxonomy of unsafe operations*. Unpublished Graduate Research Project. Daytona Beach, FL: Embry-Riddle Aeronautical University.

Primavera, L., Allison, D. and Alfonso, V. (1996). Measurement of dependent variables. In R. Franklin, D. Allison and B. Gorman (Eds.), *Design and analysis of single-case research* (pp. 41–92). Mahwah, NJ: Erlbaum.

Pounds, J., Scarborough, A. and Shappell, S. (2000). A human factors analysis of air traffic control operational errors. *Aviation Space and Environmental Medicine, 71*, 329.

Prinzo, O.V. (2001). *Data-linked pilot reply time on controller workload and communication in a simulated terminal option* (DOT/FAA/AM-01/08). Washington, DC: Federal Aviation Administration, Office of Aerospace Medicine.

Rabbe, L. (1996). *Categorizing Air Force F-16 mishaps using the taxonomy of unsafe operations.* Unpublished Graduate Research Project. Daytona Beach, FL: Embry-Riddle Aeronautical University.

Ranger, K. (1997). *Inter-rater reliability of the taxonomy of unsafe operations.* Unpublished Graduate Research Project. Daytona Beach, FL: Embry-Riddle Aeronautical University.

Rasmussen, J. (1982). Human errors: A taxonomy for describing human malfunction in industrial installations. *Journal of Occupational Accidents, 4*, 311–33.

Reason, J. (1990). *Human error.* New York: Cambridge University Press.

Reinhart, R.O. (1996). *Basic flight physiology* (2nd ed.). New York: McGraw-Hill.

Salas, E., Wilson, K., Burke, S., and Bowers, C. (2003). Myths about crew resource management training. *Ergonomics in Design, 10*(4), 20–24.

Sanders, M. & Shaw, B. (1988). *Research to determine the contribution of system factors in the occurrence of underground injury accidents.* Pittsburgh, PA: Bureau of Mines.

Sarter, N. and Woods, D.D. (1992). Pilot interaction with cockpit automation. i. operational experiences with the flight management system. *The International Journal of Aviation Psychology, 2*, 303–321.

Schmidt, J.K., Schmorrow, D. and Hardee, M. (1998). A preliminary human factors analysis of naval aviation maintenance related mishaps. *Proceedings of the 1998 Airframe/Engine Maintenance and Repair Conference (P329).* Long Beach, CA.

Senders, J.W. and Moray, N.P. (1991). *Human error: Cause, prediction and reduction.* Hillsdale, NJ: Erlbaum.

Shappell, S.A. and Wiegmann, D.A. (1995). Controlled flight into terrain: The utility of models of information processing and human error in aviation safety. *Proceedings of the Eighth Symposium on Aviation Psychology* (pp. 1300–1306). Ohio State University.

Shappell, S.A. and Wiegmann, D.A. (1996). U.S. naval aviation mishaps 1977-92: Differences between single- and dual-piloted aircraft. *Aviation, Space, and Environmental Medicine, 67*(1), 65–69.

Shappell, S.A. and Wiegmann D.A. (1997a). A human error approach to accident investigation: The taxonomy of unsafe operations. *The International Journal of Aviation Psychology, 7*(4), 269–291.

Shappell, S.A. and Wiegmann, D.A. (1997b). Why would an experienced aviator fly a perfectly good aircraft into the ground? *Proceedings of the Ninth International Symposium on Aviation Psychology* (pp. 26–32). Columbus, OH: The Ohio State University.

Shappell, S.A. and Wiegmann, D.A. (April, 1998). *Failure analysis classification system: A human factors approach to accident investigation.* SAE: Advances in Aviation Safety Conference and Exposition, Daytona Beach, FL.

Shappell, S.A. and Wiegmann, D.A. (1999). Human factors analysis of aviation accident data: Developing a needs-based, data-driven, safety program. *Proceedings of the Fourth Annual Meeting of the Human Error, Safety, and System Development Conference.* Liege, Belgium.

Shappell, S.A., Wiegmann, D.A., Fraser, J.R., Gregory, G., Kinsey, P. and Squier, H (1999). Beyond mishap rates: A human factors analysis of U.S. Navy/Marine Corps TACAIR and rotary wing mishaps using HFACS. *Aviation, Space, and Environmental Medicine,* 70, 416–417.

Shappell, S.A. and Wiegmann, D.A. (2000a). *The human factors analysis and classification system (HFACS).* (Report Number DOT/FAA/AM-00/7). Washington DC: Federal Aviation Administration.

Shappell, S.A. and Wiegmann, D.A. (2000b). Is proficiency eroding among U.S. naval aircrews? A quantitative analysis using the Human Factors Analysis and Classification System. *Proceedings of the 44th Annual Meeting of the Human Factors and Ergonomics Society,* San Diego, CA.

Shappell, S.A. and Wiegmann, D.A. (2001). Applying Reason: The human factors analysis and classification system (HFACS). *Human Factors and Aerospace Safety,* 1(1), 59–86.

Shappell, S.A. and Wiegmann, D.A. (in press). *A human error analysis of general aviation controlled flight into terrain (CFIT) accidents occurring between 1990–1998.* (Report Number DOT/FAA/AM-03/in press). Washington DC: Federal Aviation Administration.

Skinner, B.F. (1974). *About behaviorism.* New York: Vintage Books.

Suchman, E.A. (1961). *A conceptual analysis of accident phenomenon, behavioral approaches to accident research.* New York: Association for the Aid of Crippled Children.

United States Naval Safety Center (2001, November). *OPNAV Instruction 3750.6R (OPNAVINST 3750.6R).* Retrieved February 10, 2003, from http://www.safetycenter.navy.mil/instructions/aviation/opnav3750/default.htm

Walker, S. (1996). *A human factors examination of U.S. Naval controlled flight into terrain 'CFIT' Accidents.* Unpublished Graduate Research Project. Daytona Beach, FL: Embry-Riddle Aeronautical University.

Weaver, D. (October, 1971). Symptoms of operational error. *Professional Safety,* 104(2), 39–42.

Wickens, C.D. and Flach, J.M. (1988). Information processing. In E.L. Wiener and D.C. Nagel (Eds.), *Human factors in aviation* (pp. 111–55). San Diego, CA: Academic Press.

Wickens, C.D. and Hollands, J.G. (2000). *Engineering psychology and human performance* (3rd ed.). Upper Saddle River, NJ: Prentice Hall.

Wiegmann, D. and Shappell, S. (1995). Human factors in U.S. Naval aviation mishaps: An information processing approach. *Proceedings of the Eighth Symposium on Aviation Psychology.* Ohio State University.

Wiegmann, D.A. and Shappell, S.A. (1997). Human factors analysis of post-accident data: Applying theoretical taxonomies of human error. *The International Journal of Aviation Psychology*, 7(1), 67–81.

Wiegmann, D.A. and Shappell, S.A. (1999). Human error and crew resource management failures in Naval aviation mishaps: A review of U.S. Naval safety center data, 1990–96. *Aviation, Space, and Environmental Medicine*, 1147–1151.

Wiegmann, D.A., Taneja, N. and Shappell, S.A. (in press). *Analyzing medical errors and incidents: Application of the human factors analysis and classification system (HFACS)*.

Wiegmann, D.A., Rich, A. and Shappell, S.A. (2000). *Human error and accident causation theories, frameworks and analytical techniques: An annotated bibliography* (Technical Report ARL-00-12/FAA-00-7). Savoy, IL: University of Illinois, Aviation Research Lab.

Wiegmann, D., Shappell, S., Cristina, F. and Pape, A. (2000). A human factors analysis of aviation accident data: An empirical evaluation of the HFACS framework. *Aviation Space and Environmental Medicine*, 71, 328.

Wiegmann, D.A. and Shappell, S.A. (2001a). Human error perspectives in aviation. *The International of Aviation Psychology*, 11, 341–357.

Wiegmann, D.A. and Shappell, S.A. (2001b). Human error analysis of commercial aviation accidents: Application of the human factors analysis and classification system (HFACS). *Aviation, Space, and Environmental Medicine*, 72, 1006–16.

Wiegmann, D. A. and Shappell, S.A. (2001c). Assessing the reliability of the human factors analysis and classification system (HFACS) within the context of general aviation. *Aviation, Space, and Environmental Medicine*, 72(3), 266.

Wiegmann, D., Shappell, S.A., and Fraser, J.R. (2002). HFACS analysis of aviation accidents: A North American comparison. *Aviation, Space, and Environmental Medicine*, 73(3), 257.

Yacavone, D.W. (1993). Mishap trends and cause factors in Naval aviation: A review of Naval Safety Center data, 1986–90. *Aviation, Space and Environmental Medicine*, 64, 392–395.

Zotov, D. (1997). Reporting human factors accidents. *The Journal of the International Society of Air Safety Investigators*, 29, 4–20.

Index

Accident investigation, 12, 25, 48–50, 70, 80, 98, 123–4, 132–3, 137, 145, 149–50, 152, 157, 159–60
Accident rate, 2–3, 5–6, 8–11, 18, 49, 105, 117
   accident record, 5, 106, 111, 117
   airline accidents, 9, 117, 138
   aviation safety trends, 3
   commercial aviation accidents, 3, 72, 132, 137, 162
   Naval aviation accidents, 8, 11, 25, 67, 101, 107, 112, 128, 135, 137
Accident statistics, 3, 37
Active failures, 48, 50, 61, 137, 154
Adams, E. 38–9, 40, 157
ADREP-2000, 133, 157
Aeromedical perspective, 32–3, 49
Aircrew coordination, 7, 35–6, 60, 63, 111, 139
Aircrew coordination training (ACT), 111, 112, 115
Airplane manufacturers, 14
Air-traffic control (ATC), 34
Alfonso, V. 127, 159
Allison, D. 127, 159
Anastasi, A. 123, 157
Advanced qualification program (AQP), 115
Attention, 21, 23, 25, 33, 51–2, 58–9, 62, 75, 104, 111, 116–17, 130, 133, 142, 144, 146, 152, 155
Aviation industry, 2, 10, 18, 32, 34–6, 45
Aviation psychologists, 23, 34, 35

Bailey, L. 152, 157
Batt, R. 11, 159
Behavioral perspective, 30, 32
Benel, R. 25, 158
Bird, F. 38–9, 157
Boeing, 4, 117, 138, 143, 157, 158
Bowers, C. 114, 160
Brenner, C. 36–7, 157
Burke, S. 114, 160

Cacciabue, P. 133, 157
Carmines, E. 123–4, 157

Cognitive perspective, 21, 25, 49
Cohen's Kappa, 127–8
Commercial aviation, 3, 5, 9, 35–6, 72, 112, 114, 116–17, 131–2, 137, 162
Comprehensiveness, 124, 133, 137–8
Condition of operators, 56
   adverse mental states, 57, 61–2, 79–80, 89, 140
   adverse physiological states, 34, 57, 140, 141
   physical/mental limitations, 57, 59, 89, 141
Controlled flight into terrain (CFIT), 128, 138–42, 161
Crisp, V. 145, 158
Cristina, F. 162

Database, 14, 16, 133–5, 137, 141, 148, 154
Data-driven interventions, 14, 18, 104, 106, 152, 161
DeBlois, L. 157
Degani, A. 41, 43, 45, 157
Diagnosticity, 124, 138, 144
Diehl, A. 25, 157
Domino Theory, 38–9, 42

Edwards, E. 26–8, 157
Engineering, 12, 14–16, 18, 28, 35, 151–2
Environmental factors, 11, 33, 61, 145, 151
Ergonomic perspective, 26
Errors, 51–2, 140–1
   decision errors, 24, 53, 54, 62, 77, 79, 88, 95, 110–11, 118–19, 130, 133, 145, 150
   perceptual errors, 51, 54, 61, 118–19, 130, 133, 140, 142–3
   skill-based errors, 51, 53–4, 77, 79, 87, 96, 107–110, 118–21, 130, 142–4, 146, 154

Fad-driven interventions, 18
Fatigue, 12, 14–15, 18, 26, 28, 32–3, 43, 48–9, 57–8, 61, 79–80, 83, 88–9, 91, 124, 139, 156

Federal Aviation Administration (FAA), 14, 37, 81–2, 94, 96–7, 147, 154–5, 157, 159, 160–2, 164–5
Ferry, T. 43, 157
Firenze, R. 28–9, 157
Flach, J. 21–3, 161
Fleiss, J. 127, 157
Flight Safety Foundation (FSF), 4, 9, 10, 157
Ford, C. 145, 158
Four "P's" of flight deck operations, 41
Foushee, C. 34–5, 37, 60, 111, 158
Fraser, J. 8, 55, 102, 130, 158, 161–2
Fuller, R. 32, 158

Gallimore, J. 35, 37, 158
General aviation (GA), 5, 34, 43, 79, 116–121, 132, 137, 140–3, 157, 161, 162
Government Accounting Office (GAO), 114, 158
Gregory, 55, 161

Haddon, W. 36, 157–8
Hardee, M. 146, 160
Hardiman, T. 124, 133, 138, 159
Heinrich, H. 26, 38, 42, 158
Helicopter, 100, 103, 109, 115, 142
Helmreich, R. 34–5, 37, 60, 111, 158–9
HFACS, 45, 50, 62, 70–2, 75–81, 86–9, 94, 96, 98, 100–103, 105–108, 112–14, 116, 118, 121, 123–4, 127–8, 130–3, 137–8, 140, 142–3, 145–56, 161–2
Hollands, J. 62, 161
Hollnagel, E. 124, 145, 158
Human error, 2, 10–12, 15–16, 18–21, 23, 37–8, 42–5, 48–9, 57, 62–3, 70, 98, 100, 102, 105, 109, 117–18, 120–1, 123–5, 127, 133–5, 139, 145, 149–50, 152–3, 155, 160–1
Human factors, 7, 12, 15–16, 18–20, 27–8, 32, 44, 57, 123–4, 128, 134, 135, 139, 148, 150, 153–6, 158, 160–2
Huntoon, C. 61, 159

Information processing, 21–26, 59, 133–4, 155, 160–1
Instrument meteorlogical conditions (IMC), 48, 58, 97, 117, 139, 141
International Civil Aviation Organization (ICAO), 10, 15, 29, 50, 133, 158

Jack, T. 145, 158

Jensen, R. 25, 158
Johnson, W. 128, 131, 158
Jones, A. 67, 158

Kayten, P. 35, 158
Kern, T. 36, 158
Kinsey, P. 55, 161
Klein, D. 36, 157–8

Lahm, P. 1, 158
Lauber, J. 33, 158
Lautman, L. 35, 37, 158

Mechanical failures, 11–12, 14–15, 18, 45, 121, 154
Memory, 21–3, 51–2, 68, 106, 130, 144, 146
Military aviation, 4, 8, 11, 35, 50, 102, 109, 111, 132, 138, 140, 142
Moray, N. 20, 160
Morrison, D. 11, 159
Motivation, reward, and satisfaction model, 31
Muchinsky, P. 67, 159
Murray, S. 2, 26, 159

Nagel, D. 5, 8, 11, 157, 159, 161
NASA, 14, 145
National Transportation Safety Board (NTSB), 4, 10–11, 14–15, 33, 64, 72–3, 75, 77–82, 84–90, 92–8, 115–16, 137, 159
Nicogossian, A. 61, 159

O'Hare, D. 11, 23–5, 159
O'Connor, S. 124, 133, 138, 159
Orasanu, J. 36–7, 53, 159
Organizational Influences, 50, 66, 69
  operational processes, 66, 90
  organizational climate, 66, 67, 137, 158
  resource management, 66–7
Organizational perspective, 42–3, 49
Outside influence, 81–2, 97

Pape, A. 131, 162
Personal readiness, 60–1, 96, 137, 139–40
Personnel factors, 60
  crew resource management (CRM), 18, 36, 37, 48–9, 62, 65, 78, 82, 95, 111–15, 139, 141–43, 151, 158, 160, 163, 164
Petersen, D. 26, 158
Peterson, D. 30–2, 159

Peterson, L. 152, 157
Physical environment, 61
Plourde, G. 128, 130, 159
Plutarch, 3
Pool, S. 61, 159
Pounds, J. 146, 160
Preconditions for unsafe acts, 56
Prevention, 10, 12, 14, 28, 42, 49, 51, 53, 151–2, 158
Primavera, L. 159
Prinzo, O. 28, 160
Psychosocial perspective, 34, 36, 49

Rabbe, L. 128, 130, 160
Ranger, K. 128, 130, 160
Rasmussen, J. 23, 25, 51, 53, 160
Reason, J. 28, 32, 45–51, 55, 63, 70, 128, 130, 134–5, 137, 150, 154, 160–1
Reinhart, 33, 61, 158, 160
Reliability, 18, 124–8, 130–2, 134, 148, 158, 160, 162
Roos, N. 26, 158

Safety programs, 7, 16, 18–19, 32, 39, 69
Salas, E. 111, 114, 160
Sanders, M. 45, 160
Sandusky, R. 145, 158
Sarter, N. 62, 160
Scarborough, A. 146, 160
Schmidt, J. 146, 160
Schmorrow, D. 146, 160
Senders, J. 20, 160
Shappell, S. 2, 11, 18, 20, 23–4, 35–7, 50, 55, 100, 102, 107, 123–4, 128, 131–2, 134, 137–8, 140, 146, 160–2
Shaw, B. 45, 160
SHEL model, 26–28, 50
Skinner, B. 30, 161
Squier, H. 55, 161
Suchman, E. 33, 36, 157–8, 161
"Swiss cheese" model of accident causation, 47–50, 63, 70
Systems perspective, 26, 28, 30

TACAIR, 100, 103, 108–109, 130, 161

Taneja, N. 131, 162
Technological environment, 61, 82, 90

U.S. Air Force, 5, 29, 103, 106, 116, 128, 131
U.S. Navy/Marine Corps, 5–9, 11, 25, 67, 87, 98–116, 128, 132, 134–5, 137–141, 145, 152, 154, 158–9, 161–2
Unsafe acts of operators, 48, 50, 128
Unsafe supervision, 48, 63–4, 77, 80, 128
  failure to correct a known problem, 63, 65
  inadequate supervision, 63
  planned inappropriate operations, 63, 64, 80
  supervisory violations, 65
Usability, 124, 145

Validity, 102, 123–5, 132, 157
  construct validity, 123–4
  content validity, 123–4
  face validity, 123–4
Violations, 55, 57, 61, 65–6, 77–8, 87, 90, 94, 96, 101–103, 105, 120, 139–41
  exceptional violations, 55–6, 78, 87, 130, 140–1
  routine violations, 55
Visual meteorlogical conditions (VMC), 94, 117, 139, 142

Walker, S. 128, 130, 161
Weaver, D. 38–9, 161
Wickens, C. 21–3, 62, 161
Wiegmann, D. 2, 11, 18, 20, 23–4, 35–7, 50, 55, 102, 107, 123–4, 128, 131–2, 134, 137–8, 140, 146, 160–2
Wiener, E. 41, 43, 45, 157–9, 161
Wiggins, M. 11, 159
Williams, K. 152, 157, 159
Wilson, 114, 160
Woods, D. 62, 160

Yacavone, D. 5, 11, 35, 37, 112, 162

Zeller, R. 123–4, 157
Zotov, D. 23, 162